THE DARWIN READER

IN NORTON PAPERBACK EDITIONS

PETER BRENT
Charles Darwin: A Man of Enlarged Curiosity

CHARLES DARWIN
The Autobiography of Charles Darwin

CHARLES DARWIN
Darwin (Texts, Backgrounds, Contemporary Opinion, and Critical Essays: Selected and Edited by Philip Appleman)

CHARLES DARWIN
The Origin of Species (Abridged Edition, Edited by Philip Appleman)

STEPHEN JAY GOULD
Ever Since Darwin: Reflections in Natural History

STEPHEN JAY GOULD
The Flamingo's Smile: Reflections in Natural History

STEPHEN JAY GOULD
Hen's Teeth and Horse's Toes: Further Reflections in Natural History

STEPHEN JAY GOULD
The Panda's Thumb: More Reflections in Natural History

GERTRUDE HIMMELFARB
Darwin and The Darwinian Revolution

THE DARWIN READER

Edited by

MARK RIDLEY

St Catharine's College,
Cambridge University

W · W · NORTON & COMPANY
New York   London

This book was originally published in England under the
title *The Essential Darwin*

Library of Congress Cataloging-in-Publication Data

Darwin, Charles, 1809-1882.
  The Darwin reader.
  Originally published: London; Boston: Unwin Hyman, 1987
  under title: The essential Darwin.
  Bibliography: p.
  Includes index.
   1. Evolution.   2. Darwin, Charles, 1809-1882.
I. Ridley, Mark.   II. Darwin, Charles, 1809-1882.
The essential Darwin.   III. Title.
QH365.D25D37   1987   575.01′62   87-5480

ISBN 0-393-02476-8

ISBN 0-393-95673-3 {PBK.}

W. W. Norton & Company, Inc., 500 Fifth Avenue, New York, New York 10110.
W. W. Norton & Company, Ltd., 37 Great Russell Street, London WC1B 3NU.

Printed in Great Britain

1 2 3 4 5 6 7 8 9 0

Preface

Charles Darwin's complete works are like one of those multi-coursed Victorian feasts, which give us indigestion a century later even as we only read about them. Spread out over forty years, one might have got it all down without too much difficulty; but when it is assembled together the task looks hopeless, even if it is meant to be enjoyable. Where did the Victorians find their indefatigable energy of thought! And yet the effort is worthwhile. Darwin remains one of the most fascinating, the most influential, the most profoundly original and important of thinkers. And I know, from experience, how he can delight the modern reader coming to him for the first time. The beginner can face difficulties, as for instance the first two chapters of the *Origin* are by no means Darwin's most accessible piece of writing – but then (if you make it through) comes the brilliance of Chapters 3 and 4 in which he first expounds the theory of natural selection (included in this anthology as Ch. 4).

I have selected, from Darwin's nine most important books, the sections in which he explains the main ideas. I have included only a sample of the magnificent range of evidence that he collected in support of those ideas. But my aim is above all introductory, and I hope I may whet the appetite of some readers for more: after all, we do not have to finish Darwin off at one sitting. Darwin, moreover, can provide a rich entertainment even in these selected impressions and miniatures, as he describes the menus and picks out the rarer delicacies from his staggering table.

As I selected the material of this book, I received, and benefited from, the advice of Jeremy Cherfas, Michael Ghiselin, and Jonathan Hodge; they will not always agree with the final edition, but I am most grateful to them.

Mark Ridley

Contents

Contents

Contents

List of figures

Chapter one

Introduction: Charles Darwin

Although Charles Darwin is known best for his the *Origin of species*, that book is only a small part of his output and the work he put into it is only a small part of his wide-ranging inquiries. For when he published the *Origin* in 1859, at the age of 50, he was already the author of four important works and the editor of five more. He had published a theory of the formation of coral reefs which immediately replaced all previous theories and remains the accepted theory today; his treatise in four volumes on the fossil and living barnacles had re-organised that large group of animals and is still a standard reference work; the travel book now known under the title of the *Voyage of the Beagle* had been widely read; and he had published two geological books and 17 geological papers. He would go on to produce an original theory of heredity, in the two-volume *Variation under domestication*; and of sex differences, in the two-volume *Descent of man*; he would discuss the emotions at book length in the *Expression of the emotions*, and the habits and geological powers of earthworms in a charming study of the *Action of worms*. He would publish so much original work on the growth and sexual habits of flowers that, even if he had never put forward his theory of evolution, he would now be remembered as the greatest botanist of the 19th century. It is an enormous body of work, covering an enormous range of subjects. He published 19 books; his collected papers fill two modern volumes; and his correspondence will fill

1

perhaps two yards of shelf when it has all finally been printed. His biological books draw upon material from all kinds of plants and animals; he had made personal observations on many of them, and about those he had not seen he still had original ideas to contribute. He made original discoveries by close observation, in barnacles and in orchids; and by experiment, in the growth of many species of flowers; he could organise an immense range of observations made by others. But his greatest scientific achievement is as an original theorist. His theoretical ideas have been of profound historical importance; and they are so modern in conception that biologists are recurrently surprised, on their first encounter, by the relevance of Darwin's writings to their work today.

Where did Darwin's extraordinary capacity come from? He was the second son of a highly successful doctor, Robert Waring Darwin, and Susannah Wedgwood, daughter of the founder of the Wedgwood china business. The Darwins and Wedgwoods lived sufficiently near to each other in Shropshire and Staffordshire for frequent visits, and the two families were closely connected. Darwin would afterwards marry his cousin Emma Wedgwood and two of his sisters would successively marry Emma's brother and brother-in-law. Their main interest lay in 'destroying hares, pheasants partridges & woodcocks, with the aid of a double barelled gun'; but both families were intelligent and cultivated too. Charles Darwin's grandfather, Erasmus Darwin, in particular, was a man of broad-ranging intellect, the author of several biological books and botanical poems. It was for Erasmus Darwin that Coleridge (the poet, who fancied himself as an intellectual) invented the term to 'darwinize', to signify wild and fanciful speculation that Coleridge disagreed with. Erasmus Darwin had for instance 'darwinized' that living species had evolved from different forms in the past, rather than being fixed and immutable; however, although Charles and Erasmus had many shared interests, grandson probably learned little from grandfather.

'In the summer of 1818 [Darwin wrote] I went to Dr Butler's great school in Shrewsbury, and remained there for seven years.' Under Butler, the school had developed into one of the best in the country, and Butler's reforms would be an example, later in the century, to several more ancient public schools. Darwin, however, had nothing good to say about his school, or Dr Butler. When his cousin Francis Galton questioned him (among other eminent Victorians) about his life, Darwin summarily forswore his entire education. 'All I have learnt of any value has been self-taught', he said; and when he was asked whether his education had 'any particular merits?' Darwin replied 'none whatever.' He may have been a little unfair. He was an original thinker, and could hardly have been taught the particular methods he would make use of in later life: if he had been, he would not have been so original. Apart from health, high moral tone, and true religion, a liberal education would have aimed only to teach him how to think and concentrate. If Darwin wished to reject Shrewsbury as a mother, it might freely acknowledge him as a son; for in his school days, Darwin showed some of the marks of his maturer mind. He was collecting mineral specimens, flowers, insects (probably – but the main phase of his entomological enthusiasm came later), and, with his brother Erasmus, he was performing chemical experiments in a laboratory shed in the garden.

From Shewsbury, Darwin was sent to study medicine at the University of Edinburgh. And study he did; but not at medicine. It was natural history that interested him. In Edinburgh, Darwin could be found as an active member of the Plinian Society, reporting to them on the creatures of the seashore; or he would walk the shores of St Andrews as the 'zealous young friend' of the physician and zoologist Robert Edmund Grant (who, by the way, had evolutionary ideas on the 'species question'); he would attend lectures too, but in the exciting – or so he had once hoped – theatres of chemistry and mineralogy, not the dull instructions and disgusting dissection classes of the medical faculty. It became clear that Darwin would make no doctor, and his father decided that he must therefore become a clergyman. He would duly move, in 1828, to an Anglican University: he would go to Christ's College, in Cambridge.

But Cambridge was only a richer variation on the Edinburgh theme. The friendship of the 'prim and highly religious' John

Coldstream would be exchanged for that of Albert Way and a second cousin, William Darwin Fox, keen entomologists who were in no hurry to take orders. The tidal invertebrates of St Andrews would be exchanged for the beetles of the Fens; and, as Grant was left behind, the botanist John Stevens Henslow and the geologist Adam Sedgwick would step into his place. Darwin paid no more attention to theology in Cambridge than he did to medicine in Edinburgh. Of course it could not be ignored completely. He peered into the Gospels, in the original Greek; he read Locke's *Essay concerning human understanding*; but the set book that interested Darwin most was Paley's *Evidences of Christianity* and *Natural Theology*. That book expounds the 'argument from design', a proof of the existence of God. Darwin was inspired by the evidence of design in nature ('adaptation' as it is called) which Paley discussed. Later, Darwin would precisely undermine Paley's system; he would turn Paley's evidence against Paley's natural theology: for in the theory of natural selection, adaptation would have a natural, and therefore did not need a divine, origin. The *Origin of species* can be read as a systematic destruction of Paley's natural theology.

The books that most excited Darwin and his naturalist friends were the *Personal narrative* and the *Aspects of nature* of Alexander, Baron von Humboldt. 'This stirred up in me a burning zeal to add even the most humble contribution to the noble structure of Natural Science.' Humboldt was a naturalist–traveller: he had been to the tropics; he had been to Tenerife and described the colossal 'dragon tree' which stood in the garden of Dr Franqui and measured 48 feet in circumference. Darwin and his friends – Eyton, Henslow, Ransay, Way – would lay schemes to visit the Canary Isles. 'I hope you continue to fan your Canary ardor' Darwin wrote to Henslow in July 1831, 'I read and reread Humboldt, do you do the same, and I am sure nothing will prevent us from seeing the great Dragon Tree.'

But it was people, more than books, that mattered to Darwin. Henslow was much the most important. He was Darwin's mentor, companion, and host in Cambridge (Henslow kept open house once a week for his circle of scientific undergraduates); he was the main intermediary in securing the *Beagle* appointment; and he would be the friend and correspondent of Darwin to his death. Among those whom Darwin met at Henslow's house was that great scientific personage William Whewell, the Master of Trinity, 'and on several occasions I walked home with him at night.' Adam Sedgwick was another important teacher. One of perhaps the four leading geologists in the country, he would encourage Darwin, and take him on a geological expedition in Wales. Darwin had by now also obtained an introduction to the entomologist Frederick Williams Hope. Hope's main house was only six miles from Shrewsbury and he had a splendid entomological collection. Hope, too, would encourage Darwin, and take him on collecting expeditions in Wales.

Natural history merged by insensible steps in Darwin's activities with hunting and shooting. He moved from Cambridgeshire to Shropshire, to Derbyshire and Wales, and collected, hunted, and shot in all of them. He seemed to his father to be becoming a wastrel, and a series of paternal explosions sought to redirect his son back to a career. 'You care for nothing but shooting, dogs, and rat-catching, and you will be a disgrace to yourself and all your family', he had once said. The scientific establishment saw him in a different light. Darwin was the coming man, to be encouraged, and captured for science; when, in August 1831, the offer of a post for a naturalist on the research vessel HMS *Beagle* came up at short notice, Darwin was an obvious choice. Henslow was asked who should be appointed, and recommended Darwin. The offer out-trumped even the planned expedition to Tenerife: for now he would see the tropics of South America, and the South Sea Isles! He was eager to go: but his father was eager for him not to go. The voyage was a 'wild scheme', 'a useless undertaking', after which Darwin 'would never settle down to a steady life'. Josiah Wedgwood then interceded on behalf of his nephew. 'The undertaking [he wrote to Robert Darwin] would be useless as regards his profession, but looking on him as a man of enlarged curiosity, it affords him such an opportunity of seeing men and things as happens to few.' The paternal obstacle yielded and on 27 December 1831 Darwin set sail.

'The voyage of the Beagle has been by far the most important event in my life', Darwin wrote in his autobiography. 'I have always felt that I owe to the voyage the first real training or education of my mind; I was led to attend closely to several branches of natural history, and thus my powers of observation were improved, though they were always fairly developed.' The voyage lasted almost five years and took Darwin to South America, the Pacific Islands, Australia and New Zealand, the Islands of the Indian Ocean, and (briefly) to Africa. They first sailed to Tenerife 'but were prevented landing, by fears of our bringing the cholera.' Darwin never did see the great dragon tree; but now they sailed straight across the Atlantic to Brazil. There he saw a tropical forest for the first time, and was ecstatic. He wrote home 'the exquisite glorious pleasure of walking among such flowers, & such trees cannot be comprehended, but by those who have experienced it . . . I give myself great credit in not being crazy out of delight.' He could indulge his collector's instincts to the full; and as the *Beagle* called in at each port, trunks of material would be dispatched to Henslow in Cambridge.

On board the *Beagle*, Darwin was relatively isolated from the world of science. The few preceding years had been full of scientific meetings and discussion; but he would not find much of that in the company of sailors. Captain FitzRoy was a man of considerable scientific knowledge, but he was the only one. Apart from FitzRoy, the *Beagle* had a good library of scientific books, and Darwin expected scientific correspondence at each port of call. His turbulent relations with FitzRoy were undoubtedly important. Darwin had been appointed as the personal companion of that irascible man, and their relations fluctuated as one would expect them to in a small cabin for a period of five years. Darwin later reflected, in his autobiography:

Fitz-Roy's temper was a most unfortunate one. It was usually worst in the morning, and with his eagle eye he could usually detect something amiss about the ship, and was then unsparing in his blame. He was very kind to me, but was a man very difficult to live with on the intimate terms which necessarily followed from our messing ourselves in the same cabin. We had several quarrels; for instance, early in the voyage at Bahia, in Brazil, he defended and praised slavery, which I abominated, and told me that he had just visited a slave-owner, who had called up many of his slaves and asked them whether they wished to be free, and all answered 'No.' I then asked him, perhaps with a sneer, whether he thought that the answer of slaves in the presence of their master was worth anything? This made him excessively angry, and he said that as I doubted his word we could not live any longer together . . . But after a few hours Fitz-Roy showed his usual magnanimity by sending an officer to me with an apology and a request that I would continue to live with him.

His character was in several respects one of the most noble which I have known.

Of the books that Darwin read on the voyage, Lyell's *Principles of geology* influenced him the most. It persuaded him immediately. 'I am proud to remember that the first place, namely, St Jago, in the Cape de Verde archipelago, in which I geologised, convinced me of the infinite superiority of Lyell's views over those advocated in any other works known to me.' The book makes the classic case for what was derisively termed, by an opponent (William Whewell), 'uniformitarianism'. Uniformitarianism is the doctrine that geological history can be explained by such processes as can be observed in operation at present. It rules unobserved catastrophic (and, especially, supernatural) processes out of geology. In popular terms, that means it rules out the Flood; but to scientific geology it meant much more than that. For Darwin its importance was two-fold. It made him try to explain the history of the Earth by the accumulative action of natural processes; and it greatly expanded the timescale of geological history. The biblical age of the Earth was of the order of only a few thousand years; but Lyell suggested an age much greater than that. The extra years were essential in any attempt to explain geology in uniformitarian terms. In Darwin's theory of coral reefs, reefs were supposed to have grown up as the sea floor subsided to great depths; but the cumulative growth of corals, to many thousands of feet, would imply an age of more than a few thousand years. At Valdivia in Chile, Darwin saw an earthquake at first hand, and the resulting upwards movements of the Earth's crust by a few feet. But he would need more earthquakes than could easily be fitted into the biblical chronology if he was to explain how marine fossils could have been lifted from sea level to heights of thousands of feet in the Andes. The same point would apply later in this theory of evolution. The rate of evolution is so slow that we never notice it; but, if evolution continued for long enough, the modern variety of forms could have diverged from a single common ancestor.

Darwin was probably not an evolutionist during the *Beagle* voyage. Lyell wrote about the evolutionary ideas of Lamarck in the *Principles of geology*, but mainly in order to reject them. Lyell's main influence on Darwin was then geological, and the first scientific book to emerge from the *Beagle* voyage was geological too. We shall soon see why that was.

Darwin disembarked at Falmouth on 2 October 1836. By early 1837 he was living in 'dirty, odious London', which was necessary in order to arrange his collections. His father's scheme to make him a clergyman had now 'died a natural death' and Charles joined Erasmus and two of his sisters in living comfortably off the parental fortune. In London, Darwin was at the heart of Victorian science. He knew Lyell, whom 'I saw more of than any other man, before and after my marriage'; he knew Murchison; he was elected to the Geological Society of London and served as secretary from 1838 to 1842. The Geological Society was by general consent the most lively of London's scientific societies, and Darwin attended the meetings, prepared abstracts, read out reports. But his connexions extended wider. He went into 'general society'; he dined with Sydney Smith, with Macaulay, with Carlyle; he was elected a Fellow of the Royal Society, and successfully touched the Chancellor for £1000 to assist the publication of the zoology of the *Beagle* voyage. He had the country's leading zoologists and botanists at work to classify his collections. Richard Owen (who would turn sour and jealous after the publication of the *Origin*) was at work on the fossil mammals, and he was the country's first anatomist: Thomas Bell was at work on the reptiles; he was Professor of Zoology at King's College, London: John Gould, the equal of any British ornithologist, was at work on the birds, along with Eyton: Hope (and others) would describe the insect specimens: Leonard Jenyns, Henslow's brother-in-law, described the fish: and Henslow himself began the work on the plants; but the main botanical work was done later, in the 1840s, by Henslow's son-in-law and Darwin's close friend, Joseph Dalton Hooker.

While Darwin's élite corps were describing his collections, he himself had some time to work on his own ideas. His geological ideas were quite fully formed on the voyage, and with these it was mainly a matter of reading up the evidence and writing them out at full length. Darwin published at that time on several other geological topics, as might be expected from someone who was regularly encountering all manner of geological controversies. His work on the 'species question' is of greater significance. When he began his first notebook on the subject in July 1837 he had no clear theory. He was aware that species varied, for instance in relation to their environment; but that would not help with the problem posed by Paley, which had to be solved first. He wrote in his autobiography, 'It was evident that neither the action of the surrounding conditions, nor the will of the organisms [he alludes to the Lamarckian theory] could account for the innumerable cases in which organisms of every kind are beautifully adapted to their habits of life – for instance a woodpecker or a tree-frog to climb trees, or a seed for dispersal by hooks or plumes. I had always been struck by such adaptations, and until these could be explained it seemed to me almost useless to endeavour to prove by direct evidence that species had been modified.'

'In October 1838 I happened to read for amusement "Malthus on Population," and being well prepared to appreciate the struggle for existence which everywhere goes on from long-continued observation of the habits of animals and plants, it at once struck me that under these circumstances favourable variations would tend to be preserved, and unfavourable ones to be destroyed. The result of this would be the formation of new species. Here then I had at last got a theory by which to work.' Guided by this theory, he collected evidence from all appropriate sources. In 1839 he wrote a short essay, 'and this was enlarged during the summer of 1844 into one of 230 pages.'

So, in 1844, two years after Darwin moved from London to

Down House in Kent, he had written or almost completed two geological books, his travel book, and his manuscript on evolution. He had also begun observations on the fertilisation of flowers and the habits of worms. However, if we turn to the timetable of Darwin's publications, we can immediately see a difference between the geological and biological writings: his geological books came out quite quickly; but the biological books more slowly, later in his life. The pattern is probably in part the effect of environment. Darwin was actively part of the society of professional geologists, who were regularly discussing the kinds of questions Darwin also wrote about. He knew his audience, which always makes composition easier. The ideas of crustal mobility and the age of the Earth were, in theological terms, subversive; but they would scarcely be received as such while they were part of an orthodox controversy. With biological evolution the situation was quite different. Evolution had been discussed, but in sensational and anti-religious terms; it was not a controversial topic at the Linnean, Zoological and Entomological Societies in the way that Lyell's ideas were at the Geological Society. Biologists had other interests.

Then, later in 1844, Robert Chambers (anonymously) published his *Vestiges of the natural history of creation.* It was a popular work, which drew evolutionary conclusions from a most superficial and uncritical treatment of the evidence. For Darwin, *Vestiges* was a disaster. Evolution was now even more untouchable, and Darwin himself more likely to be misunderstood. The subject would now have to be treated thoroughly or not at all. His relatively light-weight essay could not be published, and he would have to keep quiet about his ideas until he had mastered all the evidence. Therefore, as we shall see in Chapter 3, when Darwin incorporated in the 1845 edition of the *Voyage of the Beagle* some of his evolutionary evidence, he would leave out its evolutionary interpretation. And in 1846 he began what turned into eight years of research on the classification of barnacles.

With the move to Down House Darwin had once more isolated himself from the scientific community. He could mature his more dangerous ideas slowly, and alone, or entrust them only to an inner circle of sympathetic friends – Lyell, Hooker, and Asa Gray, a botanist at Harvard. Although he was isolated, he was not completely cut off. He continued a large correspondence; he occasionally visited London; he went most years to the annual meeting of the British Association for the Advancement' of Science; and his scientific friends occasionally visited him. When they came, Darwin was prepared for them. After breakfast they would be 'pumped'. Hooker has described the procedure:

I had many invitations, and delightful they were. A more hospitable and more attractive home under every point of view could not be imagined—of Society there were most often Dr Falconer, Edward Forbes, Professor Bell, and Mr Waterhouse—there were long walks, romps with the children on hands and knees, music that still haunts me. Darwin's own hearty manner, hollow laugh, and thorough enjoyment of life with friends; strolls with him all together, and interviews with us one by one in his study, to discuss questions in any branch of biological or physical knowledge that we had followed; and which I at any rate always left with the feeling that I had imparted nothing and carried away more than I could stagger under. Latterly, as his health became more seriously affected, I was for days and weeks the only visitor, bringing my work with me and enjoying his society as opportunity offered. It was an established rule that he every day pumped me, as he called it, for half an hour or so after breakfast in his study, when he first brought out a heap of slips with questions botanical, geographical, &c., for me to answer, and concluded by telling me of the progress he had made in his own work, asking my opinion on various points.

As the years passed, Darwin saw less of his friends, and attended fewer scientific meetings. It was only to be expected that a man of country habits would exchange life in 'stinking, odious London' for a country house. But he would never again take up the energetic country activities he had enjoyed in his youth: he would narrow down his pursuits, and fix himself in a valetudinarian routine. To understand the change, we must turn to that fascinating, painful subject: Darwin's ill-health.

Darwin had been robustly healthy in his youth, and even in the *Beagle* voyage. To be sure, he was 'invariably sick, except on the finest of days' when on board that little ten-gun brig, as it lurched on the storms of the South Atlantic Ocean; but he was quite up to the most vigorous exploration once on shore. After his return to England, his health went into decline; and after his marriage (1839) and move to Down (1842) it broke completely. He would work for only 3–4 hours and pass the rest of the day in short walks and rests; sometimes, for periods of many weeks, he could not manage even that. In his journals, he shows an obsessive interest in the symptoms and course of his ill-health. He complained about his heart, his head, and his stomach; he visited innumerable doctors, and frequented less orthodox water-cure establishments, but no one then or since has diagnosed with any certainty what was wrong with him. Modern writers have located the cause all over the place, from a selfish scientist's Bunburyism to his (suitably psychoanalysed) relations with his father. A plausible, but not universally accepted, suggestion has been made by Professor Saul Adler. According to Adler, Darwin suffered from Chagas disease. It is indeed known that Darwin was bitten by the 'great black bug of the Pampas', the reduviid bug that transmits the disease, when he was in Argentina; moreover, Chagas disease steers a varying and unpredictable course in different cases, but often troubles the heart and intestines; and none of Darwin's doctors could have recognised the disease at the time, because it was not diagnosed until 1909.

However that may be, any real disease could be sure to flourish in the Darwin's domestic environment. For among the Wedgwoods, ill-health was a family cult. They exported the cult to the Darwins by intermarriage, and it took root in Charles's generation. His grandfather Erasmus had decayed fast, but enjoyed sound health in all of his 20 stones; he lived to the age of 70, procreating children to the end, so that his friends would knowingly remark 'too much of Venus' at his death. Erasmus's son, Charles's father, Robert Darwin did not succumb even though he married a Wedgwood; but he was a doctor. In the next generation, Erasmus and Charles first converted, and then martyred themselves, to the cult. Erasmus died a bachelor, but Charles was fortified by a double Wedgwood connexion – the decline of his health almost exactly coincides with his marriage – and piously carried the cult of ill-health into the next generation. It can be studied there in the delightful portrait *Period piece* by the artist Gwen Raverat, who was Darwin's grand-daughter. She was born three years after his death, and only had opportunity to observe his splendidly neurotic family. Six of the seven children cultivated their ill-health, and Henrietta would out-perform the excesses even of her parents. For Henrietta, 'ill-health became her profession and absorbing self-interest.' She 'was always going away to rest, in case she might be tired later in the day, or even next day'; when infection threatened, she wore an anti-cold mask of her own invention; and she was sensitive enough to drafts that she might summon a servant to 'put a silk hankerchief over her left foot as she lay on bed, because it was that much colder than her right foot.' The explanation, according to Gwen Raverat, was her grandmother. 'The trouble [she wrote] was that in my grandparents' house it was a distinction and a mournful pleasure to be ill ... It was so delightful to be pitied and nursed by my grandmother. She was extremely tender-hearted, and I have sometimes thought that she must have been rather too sorry for her family when they were unwell.' The family fully mastered the technique, as 'there was a kind of sympathetic gloating in the Darwin voices, when they said, for instance to one of us children, "And have you got a *bad* sore throat, my poor cat".' No wonder, as Gwen Raverat said, 'at Down, ill-health was considered normal.'

Ill-health only slowed down Darwin's work: it did not stop it, as the volume of his production shows. Through the 1840s and 1850s he was reading widely on the species question, collecting notes by the trunkful. 'I keep from thirty to forty large portfolios, in cabinets with labelled shelves, into which I can at once put a detached reference or memorandum. I have bought many books, and at their ends I make an index of all the facts that concern my work.' But between 1844 and the writing of the *Origin* in 1858–9 Darwin was not only collecting evidence to substantiate an already finished theory. His interest in classification, which he was working on in barnacles, led him to an important addition to the theory. In 1844, he later reflected (in his autobiography),

I overlooked one problem of great importance; and it is astonishing to me how I could have overlooked it and its solution. This problem is the tendency in organic beings descended from the same stock to diverge in character as they became modified. That they have diverged greatly is obvious from the manner in which species of all kinds can be classed under genera, genera under families, families under sub-orders and so forth . . . The solution occurred to me long after I had come to Down. The solution, as I believe, is that the modified offspring of all dominant and increasing forms tend to become adapted to many and highly diversified places in the economy of nature.

Such was Darwin's 'principle of divergence'. His theory of natural selection, in its original form, explained the adaptation of living things to their environment – which Paley had used as an argument for the existence of God. It also explained why species would change. Yet it did not explain why species had changed in such a way that they formed a diverging, tree-like branching pattern. He would now explain the branching pattern by the relatively greater struggle for existence (or competition) among more similar forms, which drives them to diverge away from each other. He would thus place more importance on intra-specific competition in the mature version of his theory. The principle of divergence is the main difference between the 1844 and 1859 forms of the theory.

After Darwin had published the *Origin*, he would shift his activities from the collection of new evidence to the publication of what he already had. He did continue to make observations, as we shall see for plants and worms; but the emphasis had shifted. The style of his writing would change at the same time. Hitherto he had taken care to make his books agreeable as well as intelligible, and enlivened them with imaginative, even poetic, passages. In the *Voyage of the Beagle*, Darwin had explained the striking difference between the faunas at equivalent latitudes in the two hemispheres with this splendid outburst:

I will recapitulate the principal facts with regard to the climate, ice-action, and organic productions of the southern hemisphere, transposing the places in imagination to Europe, with which we are so much better acquainted. Then, near Lisbon, the commonest seashells, namely, three species of Oliva, a Voluta and Terebra, would have a tropical character. In the southern provinces of France, magnificent forests, intwined by arborescent grasses and with the trees loaded with parasitical plants, would hide the face of the land. The puma and the jaguar would haunt the Pyrenees. In the latitude of Mont Blanc, but on an island as far westward as central North America, tree-ferns and parasitical Orchideae would thrive amidst the thick woods. Even as far north as central Denmark, humming birds would be seen fluttering about delicate flowers, and parrots feeding amidst the evergreen woods; and in the sea there, we should have a Voluta, and all the shells of large size and vigorous growth. Nevertheless, on some islands only 360 miles northward of our new Cape Horn in Denmark, a carcass buried in the soil would be preserved perpetually frozen. If some bold navigator attempted to penetrate northward of these islands, he would run a thousand dangers amidst gigantic icebergs, on some of which he would see great blocks of rock borne far away from their original site. Another island of large size in the latitude of Scotland, but twice as far to the west, would be "almost wholly covered with everlasting snow," and would have each bay terminated by ice-cliffs, whence great masses would be yearly detached: this island would boast only of a little moss, grass, and burnet, and a tit-lark would be its only land inhabitant. From our new Cape Horn in Denmark, a chain of mountains, scarcely half the height of the Alps, would run in a straight line due southward; and on its western flank every deep creek of the sea, or fiord, would end in "bold and astonishing glaciers." These lonely channels would frequently reverberate with the falls of ice, and so often great waves would rush along their coasts; numerous icebergs, some as tall as cathedrals, and occasionally loaded with "no inconsiderable blocks of rock," would be stranded on the outlying islets; at intervals violent earthquakes would shoot prodigious masses of ice into the waters below. Lastly, some missionaries attempting to penetrate a long arm of the sea, would behold the not lofty surrounding mountains, sending down their many grand icy streams to the sea-coast, and their progress in the boats would be checked by the innumerable floating icebergs, some small and some great; and this would have occurred on our twenty-second of June, and where the Lake of Geneva is now spread out!

The *Origin* too contains some memorable exclamations, as we shall see in the chapter on 'the struggle for existence' (pp. 81–2), and of which the final paragraphs of the book (not included in this anthology) are a better-known example. After the *Origin*, Darwin's books all have a consistently serious tone. They are still agreeable to read, occasionally for their amusing anecdotes, but mainly for the fascination of Darwin's argument. His main concessions to the reader would now be to apologise because the book was too long, or too recondite.

Through the same period Darwin was losing his own taste for literature. 'Up to the age of thirty, or beyond it, poetry of many kinds, such as the works of Milton, Gray, Byron, Wordsworth, Coleridge, and Shelley, gave me great pleasure, and even as a schoolboy I took intense delight in Shakespeare.' But later in life he found he had lost this sense. 'My mind seems to have become a kind of machine for grinding out general laws out of large collections of facts.' 'For many years I cannot endure to read a line of poetry: I have tried lately to read Shakespeare, and found it so intolerably dull that it nauseated me.' 'It is a horrid bore to feel as I constantly do, that I am a withered leaf for every subject except Science.'

Darwin was depressed also by the reception of the *Origin*. He would only have been pained when his old geological tutor Adam Sedgwick wrote from Cambridge to tell him that the *Origin*, as well as being 'utterly false and grievously mischievous', had 'greatly shocked my moral taste.' If the method of the *Origin* – i.e. the subversion of natural theology – was correct (and Sedgwick was sure it was not) 'humanity would suffer a damage that might brutalize it, and sink the human race into a lower state of degradation than any into which it has fallen since its written records tell us of its history.' That was certainly not Darwin's aim. A thick-skinned controversialist like Thomas Henry Huxley would chuckle over stuff like that, even if it were the folly of a friend; Huxley could bait bishops with relish, and liked nothing more than to crumble an opponent with superior argument. But Darwin was not a thick-skinned man. He left the controversy to his 'general agent' Huxley, and in his later books – on man, on heredity and on the habits of flowers and worms – he would write in a tone that could put no one's back up. The *Descent of man* might draw a spiteful review from the anatomist St George Mivart, but Mivart really was beyond the pale. Even Darwin had no patience with him. 'My views [Darwin wrote] have often been grossly misrepresented, bitterly opposed and ridiculed, but this has been generally done as I believe in good faith. I must however except Mr Mivart, who as an American expressed it in a letter has treated me "like a pettifogger", or as Huxley said "like an Old Bailey lawyer".' People like Mivart could never be satisfied, and therefore had to be ignored; but it was Darwin's aim 'to avoid controversies', because controversy (as Lyell advised him) 'rarely did any good and caused a miserable loss of time and temper.'

When Darwin sought to sum up the mental qualities that led to his success in science, he listed as the most important factors 'the love of science – unbounded patience in long reflecting over any subject – industry in observing and collecting facts – and a fair share of invention as well as of common sense.' He did not think of himself as clever: 'I have no great quickness of apprehension or wit which is so remarkable in some clever men, for instance, Huxley.' Nor did he claim any great aptitude for philosophy: 'my power to follow a long and purely abstract train of thought is very limited; and therefore I could never have succeeded with metaphysics or mathematics.' But he did think he possessed some power of reasoning, and 'I am superior to the common run of men in noticing things which easily escape attention, and in observing them carefully.' He also showed as much industry as possible in 'the collection of facts', by observation and reading. Indeed, Darwin's style of science is typified by profound and concentrated thought and the integration of a wide range of facts. His observation and curiosity were all-pervasive. He was thinking about everything that was going on around him, all the time. 'From my early youth I have had the strongest desire to understand or explain whatever I observed, – that is, to group all facts under some general laws.' The *Expression of the emotions* most obviously illustrates the constant activity of Darwin's mind. He reflects on the habits of his children, his pets, himself; his friends were probably unaware that, when they burst into laughter at his dinner table, their exact facial expressions were being closely scrutinised under the Darwinian microscope, and their host was asking himself why they should do *that*, of all things, to express their jovial spirits. His inquiries went beyond these everyday observations, but always he was determined to pursue the truth, even into the most minor detail. Thus when Darwin wanted to know whether worms could sense heat, he duly challenged them with a red hot poker; and then there was the question of whether worms could hear.

They took not the least notice of the shrill notes from a metal whistle, which was repeatedly sounded near them; nor did they of the deepest sounds of a bassoon. They were indifferent to shouts, if care was taken that the breath did not strike them. When placed on a table close to the keys of a piano, which was played as loudly as possible, they remained perfectly quiet.   (*Formation of vegetable mould...*)

'Worms do not possess any sense of hearing', he concluded.

The books he read underwent the same inquisition. He did not much trust his memory, but could rely on it to fetch back a crucial but dimly recalled fact. When Darwin wrote on a subject, he was the master of it: he writes as an expert and authority. But that is not what makes his writings so appealing. After all, plenty of other authors are experts. What is rare, but which marks every page of Darwin, is his determination to understand nature, his 'pure love of natural science'. His work has a transparent intellectual honesty. In 1848, Darwin wrote to his old Cambridge tutor and the patron of his *Beagle* appointment, J. S. Henslow, and speculated on why he was a scientist. 'For myself I would, however, take higher ground, for I believe there exists, & I feel within me, an instinct for truth; or knowledge or discovery, of something of the same nature as the instinct for virtue.' That is the instinct, so perfectly developed in Darwin, that gives him his eternal interest.

EDITORIAL PROCEDURE

In one sense, the 'essential' Darwin is the *Origin* and nothing more: the *Origin* was, as Darwin himself recognised, his most important book; and if he was to be remembered by one book, the *Origin* would be it. But a man of Darwin's intellect does not have to be remembered by only one book, and in selecting this 'essential' Darwin I have sought to illustrate the range of his thought. In geology, in botany, in his work on humans, Darwin can be as fascinating (and profoundly original) as when he is explicitly discussing evolution. I have not neglected evolution; but I have set it among his other leading ideas.

I have selected from nine of Darwin's books – the 'monuments of my life', as he called them. In each case I have picked the 'key' passages – that is, the passages in which Darwin expounds his main idea – and I have added some typical illustrations of the ideas in action. In the editorial introductions I have explained the context of each selection and given an outline of the whole work. I have annotated each chapter, to explain some obscure points; but the annotation is light, because most of the included passages come from Darwin's more accessible writing.

The 'key' passages are easier to identify in some cases than in others. In five of the books I have drawn upon – *Coral reefs*, the *Origin*, the *Descent of man*, *Expression of the emotions*, and the *Action of worms* – Darwin expounded, or summarised, the main ideas of the books in recognisable sections. Thus, from the *Origin*, I have lifted the main exposition of natural selection and then the summary of the evidence for evolution; from the *Descent of man*, much of the introductory chapter on the 'principles of sexual selection' and two representative passages on human evolution; from the book on worms, the summary chapter and several examples of Darwin's delightful observations (I decided to exclude the observations on habits, because – charming though they are – they are not essential to the mainly geological theme of that book). The *Expression of the emotions* has three leading ideas, and it was not difficult to extract an exposition of each of the three, and then some of the applications to real cases. From *Coral reefs*, I likewise extracted Darwin's account of the theory; but in this case I have not used any of Darwin's main evidence, partly because it would be difficult to cut out any particular 'essential' passage and partly because a more powerful test of his theory has been made in this century. I have explained it in my editorial introduction to Chapter 2. *Coral reefs* is the only representative of Darwin's extensive geological writings, which might be thought to make the anthology unbalanced; but Darwin's geological writings are often relatively technical (for reasons I have explained earlier), and many of them share the theme of crustal mobility, of which *Coral reefs* is the most famous example and therefore the best illustration.

Darwin's main work on variation and inheritance, the *Variation of animals and plants under domestication*, does not have a key passage in exactly the same sense as the five books we have just considered. However, there is an obvious choice. The penultimate chapter, in which Darwin expounds his 'provisional hypothesis of pangenesis', contains the most important theoretical idea in the book and also summarises, or at least alludes to, much of the factual material of the other chapters.

The method is less easily applied to the two botanical books I have used. In these, there is no easily identifiable key passage. The *Power of movement* does have a summary chapter, but it comes at the end of a long book and is most easily read where it was printed. Moreover, the main idea of the book – that all the movements of plants are modified versions of a single kind of movement (called 'circumnutation') – has not turned out historically to be its most influential part. But the experiments on what Darwin called heliotropism – the movement of plants towards the light – are now classical. They are part of every biological education, often in the form of a practical repeat. In this case the main modern interest of the book differs from Darwin's own emphasis, and I have preferred the former. I have therefore selected the heliotropism experiments, though I have included a short passage from another part of the book to illustrate its main theme.

In *Forms of flowers*, Darwin applies a single idea to several different species of plants. The idea is so simply stated – that 'heterostyly' in plants is an adaptation to ensure cross-fertilisation – that it did not need a separate exposition. Darwin introduces the idea instead out of a concrete observation. It comes as the

solution to a detective story – after a false start. I have chosen
Darwin's experiments on one species of cowslip; they come first
in the book, where Darwin does not pre-suppose any prior
knowledge of his idea. They are also the easiest to understand,
because the flowers of *Primula veris* come in only two forms.

The *Voyage of the Beagle*, unlike the other eight books, has no
'key' idea in the sense of the other books. It is a naturalist's travel
book. It should certainly be included in any 'essential' Darwin
because it is probably his most *read* (though not most discussed)
book and illustrates one part of his character more clearly than
any other single book: the *Beagle* book shows Darwin as a field
naturalist, of broad knowledge and acute observation. Perhaps
any part of the *Beagle* book would stand for the whole. I have
picked the Galapagos chapter. It is the most famous part of the
book (and of the voyage). It serves the double purpose here of
both illustrating the book and implicitly revealing part of the
reason why Darwin had come to accept evolution rather than the
separate creation of species.

Such has been the procedure of my selection. It does result in
some omissions, some of them large, some small. The small
omissions are few. I have not tried to *précis* Darwin's own prose –
a task that would have been as pointless as difficult. Darwin
reasoned very closely, and one of the remarkable features of his
argument is the beautiful honesty with which he considers all the
important difficulties. I have only removed a few odd passages –
usually of only a few words – in which Darwin cross-refers to
passages not included in this anthology. In the chapter on
pangenesis, I have left out longer passages of a paragraph or two. I
have also in all cases removed most of Darwin's footnotes. They
are mainly bibliographical. The short textual omissions are
indicated here only by ellipses (...). I have compiled the
anthology for the Darwin reader, not the Darwin scholar; and
while the former will not want to trip up over editorial
hieroglyphics, the latter can verify his references elsewhere.

The larger omissions are of Darwin's evidence. I have concentrated on Darwin's ideas; but of course, as well as inventing ideas,
Darwin tested them too. He typically tested them against all the

available evidence; and seems to have had a mastery of the published evidence of his time. An anthology can only introduce his ideas; it can hint at how Darwin put them to use, but it cannot show his ability to organise large arrays of evidence around them. No reader can feel just how much Darwin knew about, say, the reproductive organs and habits of animals until he has been, with Darwin, through the second part of the *Descent of man*. But there has been room to illustrate in each case the kind of evidence that Darwin used.

I have, with regret, left out one whole section of Darwin's work. It is his study of barnacles. Darwin had worked on barnacles with such singlemindedness for eight years that, when one of his children visited a neighbour, he could ask 'and when does your father do his barnacles?' They are a major part of Darwin's work and contribution to 'the noble structure of Natural Science.' But I do not think they could have been fitted into this anthology. The obviously enduring parts are descriptive and of limited interest; the less obviously enduring parts are not so readily understood as the extracts I have selected.

The anthology that follows, therefore, is mainly an anthology of Darwin's ideas. The reader will get some taste of the evidence he used to support them, but will have to return to the originals to see the full power of Darwin's arguments. The concentrated form of a selection of essential readings cannot include everything, but it can show up, in as forceful a way as a shelf of unabridged editions, the extraordinary, and fascinating, range of Darwin's thought.

Chapter two

The structure and distribution of coral reefs (1842)

At various points, scattered in the Indian and Pacific Oceans, are small coral islands called atolls. Atolls are of various shapes, but are often made up of a more-or-less complete perimeter of coral, surrounding a lagoon. They poke up in areas where the ocean is of great depth. At Keeling (now called Cocos) Island, in the Indian Ocean, 'Capt. FitzRoy at the distance of but little more than a mile from the shore sounded with a line 7200 feet long and found no bottom.' It was then known that corals could only grow in shallow water: so how could these atolls ever have come into existence? When Darwin was first thinking about the matter, the favourite hypothesis, endorsed by Lyell in his *Principles of geology* (1830), was that coral atolls grew up from volcanic craters submerged a few feet below the surface.

Darwin did not believe that theory. Submerged craters, he thought, could not possibly account for the variety of shapes of atolls. Instead, he distinguished coral islands into three main types – fringing reefs, barrier reefs, and atolls – and reasoned that a reef would gradually change from one type into the next if the reef sank relative to the sea level. A fringing reef is a coral reef around the shore of an island, a barrier reef one around an island but with an expanse of water between reef and island, and in an atoll the island has disappeared completely and only the surrounding coral remains; Darwin's figure (Fig. 2.4) shows the relation of the three types. Atolls arise by the subsidence of fringing and barrier reefs; they have nothing to do with volcanic craters.

Darwin had come to his theory, at least in outline, before he had even seen a coral reef. He wrote in his autobiography that

No other work of mine was begun in so deductive a spirit as this; for the whole work was thought out on the west coast of S. America before I had seen a true coral-reef. I had therefore only to verify and extend my views by a careful examination of living reefs. But it should be observed that I had during the previous two years been incessantly attending to the effects on the shores of S. America of the intermittent elevation of the land, together with denudation and the deposition of sediment. This necessarily led me to reflect much on the effects of subsidence, and it was easy to replace in imagination the continued deposition of sediment by the upward growth of coral. To do this was to form my theory of the formation of barrier-reefs and atolls.

As soon as Darwin saw his first atoll – at Tahiti in the Pacific in November 1835 – he was convinced. Lyell was also immediately convinced when Darwin told him the subsidence theory on returning to England in 1836–7.

Darwin began writing his book in 1838, but progress was slow. Those years were filled with activity. Besides the extensive reading required to test his theory of coral reefs, he was preparing the zoology of the *Beagle* for publication; he was secretary of the Geological Society; he was getting married. He was also reading on the 'species question'; his first essay on that subject was completed in 1842, in the same year as *Coral reefs* was finally published.

Apart from arguing the inherent probability of his coral reefs theory, how could he test it? The most convincing proof would be to demonstrate subsidence. But that could hardly be hoped for. He did collect circumstantial evidence of subsidence; but for his main test of the theory, he looked to the geographical distribution of the three types of reef. Subsidence might be expected to take place over a large area. Each type of reef should then be found in large clusters, with reefs of the same type in the same area, different types in different areas. He therefore marked on a map all the known reefs, in three colours (one for each type). The evidence fitted the prediction, and thus he could conclude his book:

Finally, when the two great types of structure, namely barrier-reefs and atolls on the one hand, and fringing reefs on the other, were laid down in colours on our map, a magnificent and harmonious picture of the movements, which the crust of the Earth has in within a late period undergone, is presented to us. We there see vast areas rising, with volcanic matter every now and then bursting forth through the vents or fissures with which they are traversed. We see other wide spaces slowly sinking without any volcanic outbursts; and we may feel sure, that this sinking must have been immense in amount as well as in area, thus to have buried over the broad face of the ocean every one of those mountains, above which atolls now stand like monuments, marking the place of their former existence.

The strongest prediction of Darwin's theory was not tested until 1952. If his theory was correct, the coral of an atoll should extend to a great depth; the living coral at the surface should be piled on many hundreds of feet of dead coral, the remains of the coral from the earlier fringing and barrier reef phases. A great depth of coral was uniquely predicted by Darwin's theory of the origin of atolls: Lyell's volcanic crater theory, for instance, would predict that coral would be found no deeper than it can grow; then, beneath the coral, would be volcanic rock. In 1952 Eniwetok atoll was drilled. Coralline limestone was found down to a depth of many thousand feet, as Darwin's theory predicted. Darwin's theory is still accepted today, with only minor modifications.

From *Coral reefs*, I have extracted the opening passage in which Darwin distinguishes the three types of reef, the summary of evidence that corals can only grow in shallow water, and then the main exposition of the theory. The exposition proceeds in three stages. Darwin first criticises the other theories of the origin of atolls; he then argues that it is likely that the bases of atolls have subsided; and he finally shows that subsidence could explain the three types of reef. I have not extracted Darwin's main test of his idea, both because it consists mainly of a large coloured map that would be difficult to reproduce and because it is in any case no longer the main evidence for the theory – the results from drilling suggest subsidence more strongly.

Introduction

The object of this volume is to describe from my own observation and the works of others, the principal kinds of coral-reefs, more especially those occurring in the open ocean, and to explain the origin of their peculiar forms. I do not here treat of the polypifers,[1] which construct these vast works, except so far as relates to their distribution, and to the conditions favourable to their vigorous growth. Without any distinct intention to classify coral-reefs, most voyagers have spoken of them under the following heads: 'lagoon-islands,' or 'atolls,' 'barrier,' or 'encircling reefs' and 'fringing,' or 'shore reefs.' The lagoon-islands have received much the most attention; and it is not surprising, for every one must be struck with astonishment, when he first beholds one of these vast rings of coral-rock, often many leagues in diameter, here and there sur-mounted by a low verdant island with dazzling white shores, bathed on the outside by the foaming breakers of the ocean, and on the inside surrounding a calm expanse of water, which, from reflection, is of a bright but pale green colour. The naturalist will feel this astonish-ment more deeply after having examined the soft and almost gelatinous bodies of these apparently insignificant creatures, and when he knows that the solid reef increases only on the outer edge, which day and night is lashed by the breakers of an ocean never at rest. Well did François Pyrard de Laval, in the year 1605, exclaim, "C'est une merueille de voir chacun de ces atollons, enuironné d'un grand banc de pierre tout autour, n'y ayant point d'artifice humain." The accompanying sketch of Whitsunday Island, in the S. Pacific, taken from Capt. Beechey's admirable Voyage, although excellent of its kind, gives but a faint idea of the singular aspect of one of these lagoon-islands. Whitsunday Island is of small size, and the whole circle has been converted into land, which is a comparatively rare circumstance. As the reef of a lagoon-island generally supports many separate small islands, the word 'island,' applied to the whole, is often the cause of confusion; hence I have invariably used in this volume the term 'atoll,' which is the name given to these circular groups of coral islets by their inhabitants in the Indian Ocean, and is synony-mous with 'lagoon-island.'

Barrier reefs, when encircling small islands, have been compara-tively little noticed by voyagers; but they well deserve attention. In their structure they are little less marvellous than atolls, and they give

Figure 2.1   Whitsunday Island, South Pacific Ocean

a singular and most picturesque character to the scenery of the islands they surround. In the accompanying sketch, taken from the voyage of the Coquille, the reef is seen from within, from one of the high peaks of the island of Bolabola. Here, as in Whitsunday island, the whole of that part of the reef which is visible is converted into land. This is a circumstance of rare occurrence; more usually a snow-white line of great breakers, with here and there an islet crowned by cocoa-nut trees, separates the smooth waters of the lagoon-like channel from the waves of the open sea. The barrier reefs of Australia and of New Caledonia, owing to their enormous dimensions, have excited much attention: in structure and form they resemble those encircling many of the smaller islands in the Pacific Ocean.

With respect to fringing, or shore reefs, there is little in their structure which needs explanation; and their name expresses their comparatively small extension. They differ from barrier-reefs in not

Figure 2.2   The island of Bolabola

lying so far from the shore, and in not having within a broad channel of deep water. Reefs also occur around submerged banks of sediment and of worndown rock; and others are scattered quite irregularly where the sea is very shallow: these in most respects are allied to those of the fringing class, but they are of comparatively little interest.

... Although this classification is useful from being obvious, and from including most of the coral reefs existing in the open sea, it admits of a more fundamental division into barrier and atoll-formed reefs on the one hand, where there is a great apparent difficulty with respect to the foundation on which they must first have grown; and into fringing reefs on the other, where, owing to the nature of the slope of the adjoining land, there is no such difficulty. ...

Several theories have been advanced to explain the origin of atolls or lagoon-islands, but scarcely one to account for barrier-reefs. From the limited depths at which reef-building polypifers can flourish, taken into consideration with certain other circumstances, we are compelled to conclude, as it will be seen, that both in atolls and barrier-reefs, the foundation on which the coral was primarily attached, has subsided; and that during this downward movement, the reefs have grown upwards. This conclusion, it will be further seen, explains most satisfactorily the outline and general form of atolls and barrier-reefs, and likewise certain peculiarities in their structure. The distribution, also, of the different kinds of coral-reefs, and their position with relation to the areas of recent elevation, and to the points subject to volcanic eruptions, fully accord with this theory of their origin.

...

On the depths at which reef-building polypifers live

... I will now describe ... the soundings off the fringing reefs of Mauritius. ... I sounded with the wide bell-shaped lead which Capt. FitzRoy used at Keeling Island, but my examination of the bottom was confined to a few miles of coast (between Port Louis and Tomb Bay) on the leeward side of the island. The edge of the reef is formed of great shapeless masses of branching Madrepores, which chiefly consist of two species,—apparently *M. corymbosa* and *pocillifera*,— mingled with a few other kinds of coral. These masses are separated from each other by the most irregular gullies and cavities, into which the lead sinks many feet. Outside this irregular border of Madrepores, the water deepens gradually to twenty fathoms, which depth generally is found at the distance of from half to three quarters of a mile from the reef. A little further out the depth is thirty fathoms, and thence the bank slopes rapidly into the depths of the ocean. This inclination is very gentle compared with that outside Keeling and other atolls, but compared with most coasts it is steep. The water was so clear outside the reef, that I could distinguish every object forming the rugged bottom. In this part, and to a depth of eight fathoms, I sounded repeatedly, and at each cast pounded the bottom with the broad lead, nevertheless the arming invariably came up perfectly clean, but deeply indented. From eight to fifteen fathoms a little calcareous sand was occasionally brought up, but more frequently the arming was simply indented. In all this space the two Madrepores above mentioned, and two species of Astræa, with rather large stars, seemed the commonest kinds; and it must be noticed that twice at the depth of 15 fathoms, the arming was marked with a clean impression of an Astræa. Besides these lithophytes, some fragments of the *Millepora alcicornis*, which occurs in the same relative position at Keeling Island, were brought up; and in the deeper parts there were large beds of a Seriatopora, different from *S. subulata*, but closely allied to it. On the beach within the reef, the rolled fragments consisted chiefly of the corals just mentioned, and of a massive Porites, like that at Keeling atoll, of a Meandrina, *Pocillopora verrucosa* and of numerous fragments of Nullipora. From fifteen to twenty fathoms the bottom was, with few exceptions, either formed of sand, or thickly covered with Seriatopora: this delicate coral seems to form at these depths extensive beds, unmingled with any other kind. At 20 fathoms, one sounding brought up a fragment of Madrepora apparently *M. pocillifera*, and I believe it is the same species (for I neglected to bring specimens from both stations) which mainly forms the upper margin of the reef; if so, it grows in depths varying from 0 to 20 fathoms. Between twenty and thirty-three fathoms I obtained several soundings and they all showed a sandy bottom, with one exception at 30 fathoms, when the arming came up scooped out, as if by the margin of a large Caryophyllia. Beyond 33 fathoms I sounded only once; and from 86 fathoms, at the distance of one mile and a third from the edge of the reef, the arming brought up calcareous sand with a pebble of volcanic rock. The circumstance of the arming having invariably come up quite clean, when sounding within a certain number of fathoms off the reefs of Mauritius and Keeling atoll (eight fathoms in the former case, and twelve in the latter), and of its having always come up (with one exception) smoothed and covered with sand, when the depth exceeded 20 fathoms, probably indicates a criterion, by which the limits of the vigorous growth of coral might in all cases be readily ascertained. I do not, however, suppose that if a vast number of soundings were obtained round these islands, the limit above assigned would be found never to vary, but I conceive the facts are sufficient to show, that the exceptions would be few. The circumstance of a *gradual* change, in the two cases, from a field of clean coral to a smooth sandy bottom, is far more important in indicating the depth at which the larger kinds of coral flourish, than almost any number of separate observations on the depth, at which certain species have been dredged up. For we can understand the gradation, only as a prolonged struggle against unfavourable conditions. If a person were to find the soil clothed with turf on the banks of a stream of water, but on going to some distance on one side of it, he observed the blades of grass growing thinner and thinner, with intervening patches of sand, until he entered a desert of sand, he would safely conclude, especially if changes of the same kind were noticed in other places, that the presence of the water was absolutely necessary to the formation of a thick bed of turf: so may we conclude, with the same feeling of certainty, that thick beds of coral are formed only at small depths beneath the surface of the sea.

I have endeavoured to collect every fact, which might either invalidate or corroborate this conclusion. Capt. Moresby, whose opportunities for observation during his survey of the Maldiva and Chagos Archipelagoes have been unrivalled, informs me, that the upper part or zone of the steep-sided reefs, on the inner and outer coasts of the atolls in both groups, invariably consists of coral, and the lower parts of sand. At seven or eight fathoms depth, the bottom is formed, as could be seen through the clear water, of great living masses of coral, which at about ten fathoms generally stand some way apart from each other, with patches of white sand between them, and at a little greater depth these patches become united into a smooth steep slope, without any coral. Capt. Moresby, also, informs me in support of his statement, that he found only decayed coral on the Padua Bank (northern part of the Laccadive group) which has an average depth between 25 and 35 fathoms, but that on some other banks in the same group with only ten or twelve fathoms water on them, (for instance, the Tillacapeni bank), the coral was living.

With regard to the coral-reefs in the Red Sea, Ehrenberg has the following passage. "The living corals do not descend there into great depths. On the edges of islets and near reefs, where the depth was small, very many lived; but we found no more even at six fathoms. The pearl-fishers at Yemen and Massaua asserted that there was no coral near the pearl-banks at nine fathoms depth, but only sand. We were not able to institute any more special researches." I am, however, assured both by Captain Moresby and Lieut. Wellstead, that in the more northern parts of the Red Sea, there are extensive beds of living coral at a depth of 25 fathoms, in which the anchors of their vessels were frequently entangled. Captain Moresby attributes the less depth, at which the corals are able to live in the places mentioned by Ehrenberg, to the greater quantity of sediment there; and the situations, where they were flourishing at the depth of 25 fathoms, were protected, and the water was extraordinarily limpid. On the leeward side of Mauritius where I found the coral growing at a somewhat greater depth than at Keeling atoll, the sea, owing apparently to its tranquil state, was likewise very clear. Within the lagoons of some of the Marshall atolls, where the water can be but little agitated, there are, according to Kotzebue, living beds of coral in 25 fathoms. From these facts, and considering the manner in which the beds of clean coral off Mauritius, Keeling Island, the Maldiva and Chagos atolls, graduated into a sandy slope, it appears very probable that the depth, at which reef-building polypifers can exist, is partly determined by the extent of inclined surface, which the currents of the sea and the recoiling waves have the power to keep free from sediment.

MM. Quoy and Gaimard believe that the growth of coral is confined within very limited depths; and they state that they never found any fragment of an Astræa, (the genus they consider most efficient in forming reefs) at a depth above 25 or 30 feet. But we have seen that in several places the bottom of the sea is paved with massive corals at more than twice this depth; and at 15 fathoms (or thrice this depth) off the reefs of Mauritius, the arming was marked with the distinct impression of a living Astræa. *Millepora alcicornis* lives in from 0 to 12 fathoms, and the genera Madrepora and Seriatopora from 0 to 20 fathoms. Capt. Moresby has given me a specimen of *Sideropora scabra* (Porites of Lamarck) brought up alive from 17 fathoms. Mr. Couthouy states that he has dredged up on the Bahama banks considerable masses of Meandrina from 16 fathoms, and he has seen this coral growing in 20 fathoms. A Caryophyllia, half an inch in diameter, was dredged up alive from 80 fathoms off Juan Fernandez (Lat. 33° S.) by Capt. P. P. King: this is the most remarkable fact with which I am acquainted, shewing the depth at which a genus of corals often found on reefs, can exist. We ought, however, to feel less surprise at this fact, as Caryophyllia alone of the lamelliform genera, ranges far beyond the tropics; it is found in Zetland[2] in Lat. 60° N. in deep water, and I procured a small species from Tierra del Fuego in Lat. 53° S. . . .

Although the limit of depth, at which each particular kind of coral ceases to exist, is far from being accurately known; yet when we bear in mind the manner in which the clumps of coral gradually became infrequent at about the same depth, and wholly disappeared at a greater depth than 20 fathoms, on the slope round Keeling atoll, on the leeward side of the Mauritius, and at rather less depth, both without and within the atolls of the Maldiva and Chagos Archipelagoes; and when we know that the reefs round these islands do not differ from other coral formations in their form and structure, we may, I think, conclude that in ordinary cases, reef-building polypifers do not flourish at greater depths than between 20 and 30 fathoms.

It has been argued that reefs may possibly rise from very great depths through the means of small corals, first making a platform for the growth of the stronger kinds. This, however, is an arbitrary supposition: it is not always remembered, that in such cases there is an antagonist power in action, namely, the decay of organic bodies, when not protected by a covering of sediment, or by their own rapid growth. We have, moreover, no right to calculate on unlimited time for the accumulation of small organic bodies into great masses. Every fact in geology proclaims that neither the land, nor the bed of the sea retain for indefinite periods the same level. As well might it be imagined that the British Seas would in time become choked up with beds of oysters, or that the numerous small corallines off the inhospitable shores of Tierra del Fuego would in time form a solid and extensive coral-reef.

Theory of the formation of the different classes of coral-reefs

The naturalists who have visited the Pacific, seem to have had their attention rivetted by the lagoon-islands, or atolls,—those singular rings of coral-land which rise abruptly out of the unfathomable ocean,—and have passed over, almost unnoticed, the scarcely less wonderful encircling barrier-reefs. The theory most generally received on the formation of atolls, is that they are based on submarine craters: but where can we find a crater of the shape of Bow atoll, which is five times as long as it is broad; or like that of Menchicoff Island, with its three loops, together sixty miles in length; or like Rimsky Korsacoff, narrow, crooked, and fifty-four miles long; or like the northern Maldiva atolls, made up of numerous ring-formed reefs, placed on the margin of a disk,—one of which disks is eighty-eight miles in length, and only from ten to twenty in breadth. It is, also, not a little improbable, that there should have existed as many craters of immense size crowded together beneath the sea, as there are now in some parts atolls. But this theory lies under a greater difficulty, as will be evident, when we consider on what foundations the atolls of the larger archipelagoes rest: nevertheless, if the rim of a crater afforded a basis at the proper depth, I am far from denying that a reef like a perfectly characterized atoll might not be formed; some such, perhaps, now exist; but I cannot believe in the possibility of the greater number having thus originated.

An earlier and better theory was proposed by Chamisso;[3] he supposes that as the more massive kinds of corals prefer the surf, the outer portions, in a reef rising from a submarine basis, would first reach the surface and consequently form a ring. But on this view it must be assumed, that in every case the basis consists of a flat bank; for if it were conically formed, like a mountainous mass, we can see no reason why the coral should spring up from the flanks, instead of from the central and highest parts: considering the number of the atolls in the Pacific and Indian Oceans, this assumption is very improbable. As the lagoons of atolls are sometimes even more than forty fathoms deep, it must, also, be assumed on this view, that at a depth at which the waves do not break, the coral grows more vigorously on the edges of a bank than on its central part: and this is an assumption without any evidence in support of it. I remarked in the third chapter, that a reef, growing on a detached bank, would tend to assume an atoll-like structure; if, therefore, corals were to grow up from a bank with a level surface some fathoms submerged, having steep sides and being situated in a deep sea, a reef not to be distinguished from an atoll, might be formed: I believe some such exist in the West Indies. But a difficulty of the same kind with that affecting the crater theory, renders, as we shall presently see, this view inapplicable to the greater number of atolls.

No theory worthy of notice has been advanced to account for those barrier-reefs, which encircle islands of moderate dimensions. The great reef which fronts the coast of Australia has been supposed, but without any special facts, to rest on the edge of a submarine precipice, extending parallel to the shore. The origin of the third class or of fringing reefs presents, I believe, scarcely any difficulty, and is simply consequent on the polypifers not growing up from great depths, and their not flourishing close to gently shelving beaches where the water is often turbid.

What cause, then, has given to atolls and barrier-reefs their characteristic forms? Let us see whether an important deduction will not follow from the consideration of these two circumstances,—first, the reef-building corals flourishing only at limited depths,—and secondly, the vastness of the areas interspersed with coral-reefs and coral-islets, none of which rise to a greater height above the level of the sea, than that attained by matter thrown up by the waves and winds. I do not make this latter statement vaguely; I have carefully sought for descriptions of every island in the inter-tropical seas; and my task has been in some degree abridged by a map of the Pacific, corrected in 1834 by M.M. D'Urville and Lottin, in which the low islands are distinguished from the high ones (even from those much less than a hundred feet in height) by being written without a capital letter. . . . I have ascertained, and chiefly from the writings of Cook, Kotzebue, Bellinghausen, Duperrey, Beechey, and Lutké, regarding the Pacific; and from Moresby with respect to the Indian Ocean, that in the following cases the term "low island" strictly means land of the height, commonly attained by matter thrown up by the winds and the waves of an open sea. If we draw a line (the plan I have always adopted) joining the external atolls of that part of the Low Archipelago in which the islands are numerous, the figure will be a pointed ellipse (reaching from Hood to Lazaref island), of which the longer axis is 840 geographical miles, and the shorter 420 miles: in this space, none of the innumerable islets, united into great rings, rise above the stated level. The Gilbert group is very narrow, and 300 miles in length. In a prolonged line from this group, at the distance of 240 miles, is the Marshall archipelago, the figure of which is an irregular square, one end being broader than the other; its length is 520 miles, with an average width of 240: these two groups together are 1040 miles in length, and all their islets are low. Between the southern end of the Gilbert and the northern end of Low Archipelago, the ocean is thinly strewed with islands, all of which, as far as I have been able to ascertain, are low: so that from nearly the southern end of the Low Archipelago to the northern end of the Marshall Archipelago, there is a narrow band of ocean, more than 4000 miles in length, containing a great number of islands, all of which are low. In the Western part of the Caroline Archipelago, there is a space of 480 miles in length, and about 100 broad, thinly interspersed with low islands. Lastly, in the Indian ocean, the archipelago of the Maldivas is 470 miles in length, and 60 in breadth; that of the Laccadives is 150 by 100 miles: as there is a low island between these two groups, they may be considered as one group of a thousand miles in length. To this may be added the Chagos group of low islands, situated 280 miles distant, in a line prolonged from the southern extremity of the Maldivas. This group, including the submerged banks, is 170 miles in length and 80 in breadth. So striking is the uniformity in direction of these three archipelagoes, all the islands of which are low, that Captain Moresby, in one of his papers, speaks of them as parts of one great chain, nearly 1500 miles long. I am, then, fully justified, in repeating, that enormous spaces, both in the Pacific and Indian oceans, are interspersed with islands, of which not one rises above that height, to which the waves and winds in an open sea can heap up matter.

On what foundations, then, have these reefs and islets of coral been constructed? A foundation must originally have been present beneath each atoll at that limited depth, which is indispensable for the first growth of the reef-building polypifers. A conjecture will perhaps be hazarded, that the requisite bases might have been afforded by the accumulation of great banks of sediment, which owing to the action of superficial currents, (aided possibly by the undulatory movement of the sea) did not quite reach the surface,—as actually appears to have been the case in some parts of the West Indian Sea. But in the form and disposition of the groups of atolls, there is nothing to countenance this notion; and the assumption without any proof, that a number of immense piles of sediment have been heaped on the floor of the great Pacific and Indian Oceans, in their central parts far remote from land, and where the dark blue colour of the limpid water bespeaks its purity, cannot for one moment be admitted.

The many widely-scattered atolls must, therefore, rest on rocky bases. But we cannot believe that the broad summit of a mountain lies buried at the depth of a few fathoms beneath every atoll, and nevertheless throughout the immense areas above-named, with not one point of rock projecting above the level of the sea; for we may judge with some accuracy of mountains beneath the sea, by those on the land; and where can we find a single chain several hundred miles in length and of considerable breadth, much less several such chains, with their many broad summits attaining the same height, within from 120 to 180 feet? If the data be thought insufficient, on which I have grounded my belief, respecting the depth at which the reef-building polypifers can exist, and it be assumed that they can flourish at a depth of even 100 fathoms, yet the weight of the above argument is but little diminished, for it is almost equally improbable, that as many submarine mountains, as there are low islands in the several great and widely-separated areas above-specified, should all rise within 600 feet of the surface of the sea and not one above it, as that they should be of the same height within the smaller limit of one or two hundred feet. So highly improbable is this supposition, that we are compelled to believe, that the bases of the many atolls did never at any one period all lie submerged within the depth of a few fathoms beneath the surface, but that they were brought into the requisite position or level, some at one period and some at another, through movements in the earth's crust. But this could not have been effected by elevation, for the belief that points so numerous and so widely-separated were successively uplifted to a certain level, but that not one point was raised above that level, is quite as improbable as the former supposition, and indeed differs little from it. It will probably occur to those who have read Ehrenberg's account of the reefs of the Red Sea, that many points in these great areas may have been elevated, but that as soon as raised, the protuberant parts were cut off by the destroying action of the waves: a moment's reflection, however, on the basin-like form of the atolls, will show that this is impossible; for the upheaval and subsequent abrasion of an island would leave a flat disk, which might become coated with coral, but not a deeply concave surface; moreover, we should expect to see, in some parts at least, the rock of the foundation brought to the surface. If, then, the foundations of the many atolls were not uplifted into the requisite position, they must of necessity have subsided into it; and this at once solves every difficulty,[4] for we may safely infer, from the facts given in the last chapter,[5] that during a gradual subsidence the corals would be favourably circumstanced for building up their solid frameworks and reaching the surface, as island after island slowly disappeared. Thus areas of immense extent in the central and most profound parts of the great oceans, might become interspersed with coral-islets, none of which would rise to a greater height than that attained by detritus heaped up by the sea, and nevertheless they might all have been formed by corals, which absolutely required for their growth a solid foundation within a few fathoms of the surface.

It would be out of place here to do more than allude to the many facts, showing that the supposition of a gradual subsidence over large areas is by no means improbable. We have the clearest proof that a movement of this kind is possible, in the upright trees buried under strata many thousand feet in thickness; we have also every reason for believing that there are now large areas gradually sinking, in the same manner as others are rising. And when we consider how many parts of the surface of the globe have been elevated within recent geological periods, we must admit that there have been subsidences on a corresponding scale, for otherwise the whole globe would have swollen. It is very remarkable that Mr. Lyell, even in the first edition of his Principles of Geology, inferred that the amount of subsidence in the Pacific must have exceeded that of elevation, from the area of land being very small relatively to the agents there tending to form it, namely, the growth of coral and volcanic action. But it will be asked, are there any direct proofs of a subsiding movement in those areas, in which subsidence will explain a phenomenom otherwise inexplicable? This, however, can hardly be expected, for it must ever be most difficult, excepting in countries long civilized, to detect a movement, the tendency of which is to conceal the part affected. In barbarous and semi-civilized nations how long might not a slow movement, even of elevation such as that now affecting Scandinavia, have escaped attention!

Mr. Williams insists strongly that the traditions of the natives, which he has taken much pains in collecting, do not indicate the appearance of any new islands: but on the theory of a gradual subsidence, all that would be apparent would be, the water sometimes encroaching slowly on the land, and the land again recovering by the accumulation of detritus its former extent, and perhaps sometimes the conversion of an atoll with coral islets on it, into a bare or into a sunken annular reef. Such changes would naturally take place at the periods when the sea rose above its usual limits, during a gale of more than ordinary strength; and the effects of the two causes would be hardly distinguishable. In Kotzebue's Voyage there are accounts of islands, both in the Caroline and Marshall Archipelagoes, which have been partly washed away during hurricanes; and Kadu, the native who was on board one of the Russian vessels, said "he saw the sea at Radack rise to the feet of the cocoa-nut trees; but it was conjured in time." A storm lately entirely swept away two of the Caroline islands, and converted them into shoals; it partly, also, destroyed two other islands. According to a tradition which was communicated to Capt. FitzRoy, it is believed in the Low Archipelago, that the arrival of the first ship caused a great inundation, which destroyed many lives. Mr. Stuchbury relates, that in 1825, the western side of Chain Atoll, in the same group, was completely devastated by a hurricane, and not less than 300 lives lost: "in this instance it was evident, even to the natives, that the hurricane alone was not sufficient to account for the violent agitation of the ocean." That considerable changes have taken place recently in some of the atolls in the Low Archipelago, appears certain from the case already given of Matilda Island: with respect to Whitsunday and Gloucester Islands in this same group, we must either attribute great inaccuracy to their discoverer, the famous circumnavigator Wallis, or believe that they have undergone a considerable change in the period of fifty-nine years, between his voyage and that of Capt. Beechey's. Whitsunday Island is described by Wallis as "about four miles long, and three wide," now it is only one mile and a half long. The appearance of Gloucester Island, in Capt. Beechey's words, "has been accurately described by its discoverer, but its present form and extent differ materially." Blenheim reef, in the Chagos group, consists of a water-washed annular reef, thirteen miles in circumference, surrounding a lagoon ten fathoms deep; on its surface there were a few worn patches of conglomerate coral-rock, of about the size of hovels; and these Capt. Moresby considered, as being, without doubt, the last remnants of islets; so that here an atoll has been converted into an atoll-formed reef. The inhabitants of the Maldiva Archipelago, as long ago as 1605, declared, "that the high tides and violent currents were always diminishing the number of the islands:" and I have already shown, on the authority of Capt. Moresby, that the work of destruction is still in progress; but that on the other hand the first formation of some islets is known to the present inhabitants. In such cases, it would be exceedingly difficult to detect a gradual subsidence of the foundation, on which these mutable structures rest.

Some of the archipelagoes of low coral-islands are subject to earthquakes: Capt. Moresby informs me that they are frequent, though not very strong, in the Chagos group, which occupies a very central position in the Indian ocean, and is far from any land not of coral formation. One of the islands in this group was formerly covered by a bed of mould, which, after an earthquake, disappeared, and was believed by the residents to have been washed by the rain through the broken masses of underlying rock: the island was thus rendered unproductive. Chamisso states, that earthquakes are felt in the Marshall atolls, which are far from any high land, and likewise in the islands of the Caroline Archipelago. On one of the latter, namely Oulleay atoll, Admiral Lutké, as he had the kindness to inform me, observed several straight fissures about a foot in width, running for some hundred yards obliquely across the whole width of the reef. Fissures indicate a stretching of the earth's crust, and, therefore, probably changes in its level; but these coral-islands, which have been shaken and fissured, certainly have not been elevated, and, therefore, probably they have subsided. . . .

The facts stand thus;—there are many large tracts of ocean, without any high land, interspersed with reefs and islets, formed by the growth of those kinds of corals, which cannot live at great depths; and the existence of these reefs and low islets, in such numbers and at such distant points, is quite inexplicable, excepting on the theory, that the bases on which the reefs first became attached, slowly and successively sank beneath the level of the sea, whilst the corals continued to grow upwards. No positive facts are opposed to this view, and some general considerations render it probable. There is evidence of change in form, whether or not from subsidence, on some of these coral-islands; and there is evidence of subterranean disturbances beneath them. Will then the theory, to which we have thus been led, solve the curious problem,—what has given to each class of reef its peculiar form?

Let us in imagination place within one of the subsiding areas, an island surrounded by a "fringing reef,"—that kind, which alone offers no difficulty in the explanation of its origin. Let the unbroken lines and the oblique shading in the woodcut [Figure 2.3] represent a vertical section through such an island; and the horizontal shading will represent the section of the reef. Now, as the island sinks down, either a few feet at a time or quite insensibly, we may safely infer from what we know of the conditions favourable to the growth of coral, that the living masses bathed by the surf on the margin of the reef, will soon regain the surface. The water, however, will encroach, little by little, on the shore, the island becoming lower and smaller, and the

38

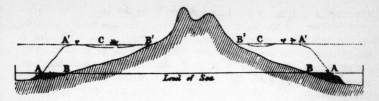

Figure 2.3   Types of coral-reef

AA – Outer edge of the reef at the level of the sea.

BB – Shores of the island.

A'A' – Outer edge of the reef, after its upward growth during a period of subsidence.

CC – The lagoon-channel between the reef and the shores of the now encircled island.

B'B' – The shores of the encircled island.

   N.B. In this, and the following wood-cut [Fig. 2.4], the subsidence of the land could only be represented by an apparent rise in the level of the sea.

space between the edge of the reef and the beach proportionally broader. A section of the reef and island in this state, after a subsidence of several hundred feet, is given by the dotted lines: coral-islets are supposed to have been formed on the new reef, and a ship is anchored in the lagoon-channel. This section is in every respect that of an encircling barrier-reef; it is, in fact, a section taken[6] E. and W. through the highest point of the encircled island of Bolabola. . . . The same section is more clearly shown in the following woodcut [Figure 2.4] by the unbroken lines. The width of the reef, and its slope both on the outer and inner side, will have been determined by the growing powers of the coral, under the conditions, (for instance the force of the breakers and of the currents) to which it has been exposed; and the lagoon-channel will be deeper or shallower, in proportion to the growth of the delicately branched corals within the reef, and to the accumulation of sediment, relatively, also, to the rate of subsidence and the length of the intervening stationary periods.

   It is evident in this section, that a line drawn perpendicularly down from the outer edge of the new reef to the foundation of solid rock, exceeds by as many feet as there have been feet of subsidence, that small limit of depth at which the effective polypifers can live,—the corals having grown up, as the whole sank down, from a basis formed of other corals and their consolidated fragments. Thus the difficulty on this head, which before seemed so great, disappears.

   As the space between the reef and the subsiding shore continued to increase in breadth and depth, and as the injurious effects of the sediment and fresh water borne down from the land were consequently lessened, the greater number of the channels, with which the reef in its fringing state must have been breached, especially those which fronted the smaller streams, will have become choked up by the growth of coral: on the windward side of the reef, where the coral grows most vigorously, the breaches will probably have first been closed. In barrier-reefs, therefore, the breaches kept open by draining the tidal waters of the lagoon-channel, will generally be placed on the leeward side, and they will still face the mouths of the larger streams, although removed beyond the influence of their sediment and fresh-water;—and this, it has been shown, is commonly the case.

Referring to the following diagram, in which the newly-formed barrier-reef is represented by unbroken lines, instead of by dots as in the former woodcut, let the work of subsidence go on, and the doubly-pointed hill will form two small islands (or more, according to the number of the hills) included within one annular reef. Let the island continue subsiding, and the coral-reef will continue growing up on its own foundation, whilst the water gains inch by inch on the land, until the last and highest pinnacle is covered, and there remains a perfect atoll. A vertical section of this atoll is shown in the woodcut by the dotted lines; a ship is anchored in its lagoon, but islets are not supposed yet to have been formed on the reef. The depth of the lagoon and the width and slope of the reef, will depend on the

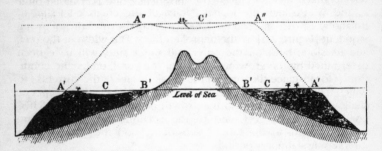

Figure 2.4 Types of coral-reef

A′A′ – Outer edges of the barrier-reef at the level of the sea. The cocoa-nut trees represent coral-islets formed on the reef.

CC – The lagoon-channel.

B′B′ – The shores of the island, generally formed of low alluvial land and of coral detritus from the lagoon-channel.

A″A″ – The outer edges of the reef now forming an atoll.

C′ – The lagoon of the newly-formed atoll. According to the scale, the depth of the lagoon and of the lagoon-channel is exaggerated.

circumstances just referred to under barrier-reefs. Any further subsidence will produce no change in the atoll, except perhaps a diminution in its size, from the reef not growing vertically upwards; but should the currents of the sea act violently on it, and should the corals perish on part or on the whole of its margin, changes would result during subsidence which will be presently noticed. I may here observe, that a bank either of rock or of hardened sediment, level with the surface of the sea, and fringed with living coral, would (if not so small as to allow the central space to be quickly filled up with detritus) by subsidence be converted immediately into an atoll, without passing, as in the case of a reef fringing the shore of an island, through the intermediate form of a barrier-reef. If such a bank lay a few fathoms submerged, the simple growth of the coral . . . without the aid of subsidence, would produce a structure scarcely to be distinguished from a true atoll; for in all cases the corals on the outer margin of a reef, from having space and being freely exposed to the open sea, will grow vigorously and tend to form a continuous ring, whilst the growth of the less massive kinds on the central expanse, will be checked by the sediment formed there, and by that washed inwards by the breakers; and as the space becomes shallower, their growth will, also, be checked by the impurities of the water, and probably by the small amount of food brought by the enfeebled currents, in proportion to the surface of the living reefs studded with innumerable craving mouths: the subsidence of a reef based on a bank of this kind, would give depth to its central expanse or lagoon, steepness to its flanks, and through the free growth of the coral, symmetry to its outline:—I may here repeat that the larger groups of atolls in the Pacific and Indian oceans cannot be supposed to be founded on banks of this nature.

If, instead of the island in the diagram, the shore of a continent fringed by a reef had subsided, a great barrier-reef, like that on the N.E. coast of Australia, would have necessarily resulted; and it would have been separated from the main land by a deep-water channel, broad in proportion to the amount of subsidence, and to the less or greater inclination of the neighbouring coast-land. The effect of the continued subsidence of a great barrier-reef of this kind, and its probable conversion into a chain of separate atolls, will be noticed, when we discuss the apparent progressive disseverment of the larger Maldiva atolls.

We now are able to perceive that the close similarity in form, dimensions, structure, and relative position (which latter point will hereafter be more fully noticed) between fringing and encircling barrier-reefs, and between these latter and atolls, is the necessary result of the transformation, during subsidence, of the one class into the other. On this view, the three classes of reefs ought to graduate into each other. Reefs having an intermediate character between those of the fringing and barrier classes do exist; for instance, on the S.W. coast of Madagascar, a reef extends for several miles, within which there is a broad channel from seven to eight fathoms deep, but the sea does not deepen abruptly outside the reef. Such cases, however, are open to some doubts, for an old fringing reef, which had extended itself a little on a basis of its own formation, would hardly be distinguishable from a barrier-reef, produced by a small amount of subsidence, and with its lagoon-channel nearly filled up with sediment during a long stationary period. Between barrier-reefs, encircling either one lofty island or several small low ones, and atolls including a mere expanse of water, a striking series can be shown.

'In proof of this' Darwin continues 'I need only refer' to the illustrations published in the original volume, which show 'more plainly to the eye, than any description could to the ear' just such a striking series. He then considers how the same theory can account for the full diversity of coral reefs, as well as the three main types, and concludes as follows.

In this chapter it has, I think, been shown, that the theory of subsidence, which we were compelled to receive from the necessity of giving to the corals, in certain large areas, foundations at the requisite depth, explains both the normal structure and the less regular forms of those two great classes of reefs, which have justly excited the astonishment of all persons who have sailed through the Pacific and Indian oceans.

Notes

1 Which we now call corals.
2 Now called Shetland.
3 Adelbert von Chamisso, 1781–1838. Franco-German author and naturalist. Author of the story *Peter Schlemihl*.
4 [Darwin's note] The additional difficulty on the crater hypothesis before alluded to, will now be evident; for on this view the volcanic action must be supposed to have formed within the areas specified a vast number of craters, all rising within a few fathoms of the surface, and not one above it. The supposition that the craters were at different times upraised above the surface, and were there abraded by the surf and subsequently coated by corals, is subject to nearly the same objections with those given at the bottom of the last page; but I consider it superfluous to detail all the arguments opposed to such a notion. Chamisso's

theory, from assuming the existence of so many banks, all lying at the proper depth beneath the water, is also vitally defective. The same observation applies to an hypothesis of Lieut. Nelson's, who supposes that the ring-formed structure is caused by a greater number of germs of corals becoming attached to the declivity, than to the central plateau of a submarine bank: it likewise applies to the notion formerly entertained, that lagoon-islands owe their peculiar form to the instinctive tendencies of the polypifers. According to this later view, the corals on the outer margin of the reef instinctively expose themselves to the surf in order to afford protection to corals living in the lagoon, which belong to other genera, and to other families!

5   Of which most of the third section – 'on the depth at which corals live' – has been reproduced here.

6   [Darwin's note]   The section has been made from the chart given in the Atlas of the Voyage of the *Coquille*. The scale is [*c*.] ·57 of an inch to a mile. The height of the island, according to M. Lesson, is 4026 feet. The deepest part of the lagoon-chanel is 162 feet; its depth is exaggerated in the woodcut for the sake of clearness.

─────── *Chapter three* ───────
The voyage of the Beagle
(1845, 1st edn 1839)

The *Voyage of the Beagle*, as it is now usually called, is Darwin's most enduringly popular book. It is easily readable, and its Victorian sense of purpose is constantly enlivened by the author's own youthful charm, his occasional poetic passages, his consistently fascinating accounts of the inhabitants, the animals, the plants, the geology of exotic places. Henslow had assured Darwin, when the *Beagle* appointment was still only an offer, 'I consider you to be the best qualified person I know of who is likely to undertake such a situation.' Henslow's judgement is confirmed by the book, which amply demonstrates Darwin's abilities as a naturalist, his powers of observation and knowledge of living things.

The book is a chronological narrative of the five year voyage. It describes in turn Brazil and Argentina, the pampas, the Falkland Islands ('these miserable islands'), Tierra del Fuego, Chile, the Galapagos and Pacific islands, New Zealand, Australia, and the islands of the Indian Ocean. The tone varies with Darwin's fluctuating moods, but any part could illustrate the whole. I have picked the most famous chapter, the one on that 'little world within itself', the Galapagos Islands.

Darwin started work on the book soon after settling in London. Work was slow, for, as Darwin wrote to Henslow, 'it is an awful thing to say to oneself, every fool and every clever man in England, if he chooses, may make as many ill-natured remarks as he likes on this unfortunate sentence.' He had completed it in June 1837, when his evolutionary ideas were scarcely formulated. It was to be published as part of a full three-volume account of the voyage, and publication was delayed while the other authors dithered; but it finally came out in 1839 under the title *Journal of researches*. . . . Initially it was buried in a three-volume official report, but it was soon published separately. Darwin produced a revised edition in 1845 and it is from that edition that I have taken the extract. The book was highly successful. 'The success of this my first literary child always tickles my vanity more than that of any of my other books. Even to this day [sometime in 1876–81] it sells steadily in England and the United States.' The edition of 1839 contained little hint of Darwin's evolutionary views; but between writing the editions of 1839 and 1845 he had become convinced of evolution, and the extensive changes for the second edition can, at least retrospectively, be seen to reflect that. The changes particularly came in the Galapagos chapter, which grew by more than half. His subsequent study of the collections he made on those islands had played no small part in confirming his evolutionary ideas, and in the revised edition he described the finches in much more detail and included Hooker's analysis of the Galapagos flora. The high degree of endemism, which is illustrated by the table (p. 65) for plants, is particularly suggestive of their evolutionary origin on the islands; so too is the presence of many closely related species within several groups – the finches are the best known case, but the same is true of tortoises, lizards, and many plants, as Darwin shows. The fact could most easily be explained by the evolution of the related species from a common ancestor on the islands. But Darwin's evolutionary ideas have to be read into the chapter (see especially p. 66) because he keeps them wholly implicit. I have extracted the text of the revised edition because, as well as illustrating this most delightful book, it suggests the kinds of evidence that convinced Darwin of evolution.

Galapagos archipelago

September 15th.—This archipelago consists of ten principal islands, of which five exceed the others in size. They are situated under the Equator, and between five and six hundred miles westward of the coast of America. They are all formed of volcanic rocks; a few fragments of granite curiously glazed and altered by the heat, can hardly be considered as an exception. Some of the craters, surmounting the larger islands, are of immense size, and they rise to a height of between three and four thousand feet. Their flanks are studded by innumerable smaller orifices. I scarcely hesitate to affirm, that there must be in the whole archipelago at least two thousand craters. These consist either of lava and scoriæ, or of finely-stratified, sandstone-like tuff. Most of the latter are beautifully symmetrical; they owe their origin to eruptions of volcanic mud without any lava: it is a remarkable circumstance that every one of the twenty-eight tuff-craters which were examined, had their southern sides either much lower than the other sides, or quite broken down and removed. As all these craters have apparently been formed when standing in the sea, and as the waves from the trade-wind and the swell from the open Pacific here unite their forces on the southern coasts of all the islands, this singular uniformity in the broken state of the craters, composed of the soft and yielding tuff, is easily explained.

Considering that these islands are placed directly under the equator, the climate is far from being excessively hot; this seems chiefly caused by the singularly low temperature of the surrounding water, brought here by the great southern Polar current. Excepting during one short season, very little rain falls, and even then it is irregular; but the clouds generally hang low. Hence, whilst the lower parts of the island are very sterile, the upper parts, at a height of a thousand feet and upwards, possess a damp climate and a tolerably luxuriant vegetation. This is especially the case on the windward

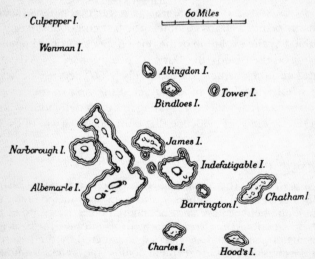

Figure 3.1   Map of Galapagos archipelago

sides of the islands, which first receive and condense the moisture from the atmosphere.

In the morning (17th) we landed on Chatham Island, which, like the others, rises with a tame and rounded outline, broken here and there by scattered hillocks, the remains of former craters. Nothing could be less inviting than the first appearance. A broken field of black basaltic lava, thrown into the most rugged waves, and crossed by great fissures, is every where covered by stunted, sunburnt brush-wood, which shows little signs of life. The dry and parched surface, being heated by the noonday sun, gave to the air a close and sultry feeling, like that from a stove: we fancied even that the bushes smelt unpleasantly. Although I diligently tried to collect as many plants as possible, I succeeded in getting very few; and such wretched-looking little weeds would have better become an arctic than an equatorial Flora. The brushwood appears, from a short distance, as leafless as our trees during winter; and it was some time before I discovered that not only almost every plant was now in full leaf, but that the greater number were in flower. The commonest bush is one of the Euphorbiaceæ: an acacia and a great odd-looking cactus are the only trees which afford any shade. After the season of heavy rains, the islands are said to appear for a short time partially green. The volcanic island of Fernando Noronha,[1] placed in many respects under nearly similar conditions, is the only other country where I have seen a vegetation at all like this of the Galapagos islands.

The *Beagle* sailed round Chatham Island, and anchored in several bays. One night I slept on shore on a part of the island, where black truncated cones were extraordinarily numerous: from one small eminence I counted sixty of them, all surmounted by craters more or less perfect. The greater number consisted merely of a ring of red scoriæ or slags, cemented together: and their height above the plain of lava was not more than from fifty to a hundred feet: none had been very lately active. The entire surface of this part of the island seems to have been permeated, like a sieve, by the subterranean vapours: here and there the lava, whilst soft, has been blown into great bubbles; and in other parts, the tops of caverns similarly formed have fallen in, leaving circular pits with steep sides. From the regular form of the many craters, they gave to the country an artificial appearance, which vividly reminded me of those parts of Staffordshire, where the great iron-foundries are most numerous. The day was glowing hot, and the scrambling over the rough surface and through the intricate thickets, was very fatiguing; but I was well repaid by the strange Cyclopean scene. As I was walking along I met two large tortoises, each of which must have weighed at least two hundred pounds: one was eating a piece of cactus, and as I approached, it stared at me and slowly stalked away; the other gave a deep hiss, and drew in its head. These huge reptiles, surrounded by the black lava, the leafless shrubs, and large cacti, seemed to my fancy like some antediluvian animals. The few dull-coloured birds cared no more for me, than they did for the great tortoises.

23rd.—The *Beagle* proceeded to Charles Island. This archipelago has long been frequented, first by the Bucaniers, and latterly by whalers, but it is only within the last six years, that a small colony has been established here. The inhabitants are between two and three hundred in number: they are nearly all people of colour, who have been banished for political crimes from the Republic of the Equator, of which Quito is the capital. The settlement is placed about four and a half miles inland, and at a height probably of a thousand feet. In the first part of the road we passed through leafless thickets, as in Chatham Island. Higher up, the woods gradually became greener: and as soon as we crossed the ridge of the island, we were cooled by a fine southerly breeze, and our sight refreshed by a green and thriving vegetation. In this upper region coarse grasses and ferns abound; but there are no tree-ferns: I saw nowhere any member of the Palm family, which is the more singular, as 360 miles northward, Cocos Island takes its name from the number of cocoa-nuts. The houses are irregularly scattered over a flat space of ground, which is cultivated with sweet potatoes and bananas. It will not easily be imagined how pleasant the sight of black mud was to us, after having been so long accustomed to the parched soil of Peru and northern Chile. The inhabitants, although complaining of poverty, obtain, without much trouble, the means of subsistence. In the woods there are many wild pigs and goats; but the staple article of animal food is supplied by the tortoises. Their numbers have of course been greatly reduced in this island, but the people yet count on two days' hunting giving them food for the rest of the week. It is said that formerly single vessels have taken away as many as seven hundred, and that the ship's company of a frigate some years since brought down in one day two hundred tortoises to the beach.

September 29th.—We doubled the south-west extremity of Albemarle Island, and the next day were nearly becalmed between it and Narborough Island. Both are covered with immense deluges of black naked lava, which have flowed either over the rims of the great caldrons, like pitch over the rim of a pot in which it has been boiled, or have burst forth from smaller orifices on the flanks; in their descent they have spread over miles of the sea-coast. On both of these islands, eruptions are known to have taken place; and in Albemarle, we saw a small jet of smoke curling from the summit of one of the great craters. In the evening we anchored in Bank's Cove, in Albemarle Island. The next morning I went out walking. To the south of the broken tuff-crater, in which the *Beagle* was anchored, there was another beautifully symmetrical one of an elliptic form; its longer axis was a little less than a mile, and its depth about 500 feet. At its bottom there was a shallow lake, in the middle of which a tiny crater formed an islet. The day was overpoweringly hot, and the lake looked clear and blue: I hurried down the cindery slope, and choked with dust eagerly tasted the water—but, to my sorrow, I found it salt as brine.

The rocks on the coast abounded with great black lizards, between three and four feet long; and on the hills, an ugly yellowish-brown species was equally common. We saw many of this latter kind, some clumsily running out of our way, and others shuffling into their burrows. I shall presently describe in more detail the habits of both these reptiles. The whole of this northern part of Albemarle Island is miserably sterile.

October 8th.—We arrived at James Island; this island, as well as Charles Island, were long since thus named after our kings of the Stuart line. Mr. Bynoe, myself, and our servants were left here for a week, with provisions and a tent, whilst the *Beagle* went for water. We found here a party of Spaniards, who had been sent from Charles Island to dry fish, and to salt tortoise-meat. About six miles inland, and at the height of nearly 2000 feet, a hovel had been built in which two men lived, who were employed in catching tortoises, whilst the others were fishing on the coast. I paid this party two visits, and slept there one night. As in the other islands, the lower region was covered by nearly leafless bushes, but the trees were here of a larger growth than elsewhere, several being two feet and some even two feet nine inches in diameter. The upper region being kept damp by the clouds, supports a green and flourishing vegetation. So damp was the ground, that there were large beds of a coarse cyperus, in which great numbers of a very small water-rail lived and bred. While staying in this upper region, we lived entirely upon tortoise-meat: the breast-plate roasted (as the Gauchos do *carne con cuero*), with the flesh on it, is very good; and the young tortoises make excellent soup; but otherwise the meat to my taste is indifferent.

One day we accompanied a party of the Spaniards in their whale-boat to a salina, or lake from which salt is procured. After landing, we had a very rough walk over a rugged field of recent lava, which has almost surrounded a tuff-crater, at the bottom of which the salt-lake lies. The water is only three or four inches deep, and rests on a layer of beautifully crystallized, white salt. The lake is quite circular, and is fringed with a border of bright green succulent plants; the almost precipitous walls of the crater are clothed with wood, so that the scene was altogether both picturesque and curious. A few years since, the sailors belonging to a sealing-vessel murdered their captain in this quiet spot; and we saw his skull lying among the bushes.

During the greater part of our stay of a week, the sky was cloudless, and if the trade-wind failed for an hour, the heat became very oppressive. On two days, the thermometer within the tent stood for some hours at 93°; but in the open air, in the wind and sun, at only 85°. The sand was extremely hot; the thermometer placed in some of a brown colour immediately rose to 137°, and how much above that it would have risen, I do not know, for it was not graduated any higher. The black sand felt much hotter, so that even in thick boots it was quite disagreeable to walk over it.

The natural history of these islands is eminently curious, and well deserves attention. Most of the organic productions are aboriginal creations, found nowhere else; there is even a difference between the inhabitants of the different islands; yet all show a marked relationship with those of America, though separated from that continent by an open space of ocean, between 500 and 600 miles in width. The archipelago is a little world within itself, or rather a satellite attached to America, whence it has derived a few stray colonists, and has received the general character of its indigenous productions. Considering the small size of these islands, we feel the more astonished at the number of their aboriginal beings, and at their confined range. Seeing every height crowned with its crater, and the boundaries of most of the lava-streams still distinct, we are led to believe that within a period, geologically recent, the unbroken ocean was here spread out. Hence, in both space and time, we seem to be brought somewhat near to that great fact—that mystery of mysteries—the first appearance of new beings on this earth.

Of terrestrial mammals, there is only one which must be considered as indigenous, namely, a mouse (*Mus galapagoensis*), and this is confined, as far as I could ascertain, to Chatham Island, the most easterly island of the group. It belongs, as I am informed by Mr. Waterhouse, to a division of the family of mice characteristic of America. At James Island, there is a rat sufficiently distinct from the common kind to have been named and described by Mr. Waterhouse; but as it belongs to the old-world division of the family, and as this island has been frequented by ships for the last hundred and fifty years, I can hardly doubt that this rat is merely a variety, produced by the new and peculiar climate, food, and soil, to which it has been subjected. Although no one has a right to speculate without distinct facts, yet even with respect to the Chatham Island mouse, it should be borne in mind, that it may possibly be an American species imported here; for I have seen, in a most unfrequented part of the Pampas, a

50

native mouse living in the roof of a newly-built hovel, and therefore its transportation in a vessel is not improbable: analogous facts have been observed by Dr. Richardson in North America.

Of land-birds I obtained twenty-six kinds, all peculiar to the group and found nowhere else, with the exception of one lark-like finch from North America (*Dolichonyx oryzivorus*), which ranges on that continent as far north as 54°, and generally frequents marshes. The other twenty-five birds consist, firstly, of a hawk, curiously intermediate in structure between a buzzard and the American group of carrion-feeding Polybori; and with these latter birds it agrees most closely in every habit and even tone of voice. Secondly, there are two owls, representing the short-eared and white barn-owls of Europe. Thirdly, a wren, three tyrant fly-catchers (two of them species of *Pyrocephalus*, one or both of which would be ranked by some ornithologists as only varieties), and a dove—all analogous to, but distinct from, American species. Fourthly, a swallow, which though differing from the *Progne purpurea* of both Americas, only in being rather duller coloured, smaller, and slenderer, is considered by Mr. Gould as specifically distinct. Fifthly, there are three species of mocking-thrush—a form highly characteristic of America. The remaining land-birds form a most singular group of finches, related to each other in the structure of their beaks, short tails, form of body, and plumage: there are thirteen species, which Mr. Gould has divided into four sub-groups. All these species are peculiar to this archipelago; and so is the whole group, with the exception of one species of the sub-group *Cactornis*, lately brought from Bow Island, in the Low Archipelago. Of *Cactornis*, the two species may be often seen climbing about the flowers of the great cactus-trees; but all the other species of this group of finches, mingled together in flocks, feed on the dry and sterile ground of the lower districts. The males of all, or certainly of the greater number, are jet black; and the females (with perhaps one or two exceptions) are brown. The most curious fact is the perfect gradation in the size of the beaks in the different species of *Geospiza*, from one as large as that of a haw-finch to that of a chaffinch, and (if Mr. Gould is right in including his sub-group, *Certhidea*, in the main group), even to that of a warbler. The largest beak in the genus *Geospiza* is shown in Fig. 1 [numbers refer to parts of Fig. 3.2], and the smallest in Fig. 3; but instead of there being only one intermediate species, with a beak of the size shown in Fig. 2, there are no less than six species with insensibly graduated beaks. The beak of the sub-group *Certhidea*, is shown in Fig. 4. The beak of *Cactornis* is somewhat like that of a starling; and that of the fourth sub-group, *Camarhynchus*, is slightly parrot-shaped. Seeing this gradation and diversity of structure in one small, intimately related

51

The voyage of the Beagle

group of birds, one might really fancy that from an original paucity of birds in this archipelago, one species had been taken and modified for different ends. In a like manner it might be fancied that a bird originally a buzzard, had been induced here to undertake the office of the carrion-feeding Polybori of the American continent.

Of waders and water-birds I was able to get only eleven kinds, and of these only three (including a rail confined to the damp summits of the islands) are new species. Considering the wandering habits of the gulls, I was surprised to find that the species inhabiting these islands is peculiar, but allied to one from the southern parts of South America. The far greater peculiarity of the land-birds, namely, twenty-five out of twenty-six being new species or at least new races, compared with the waders and web-footed birds, is in accordance with the greater range which these latter orders have in all parts of the world. We shall hereafter see this law of aquatic forms, whether marine or fresh-water, being less peculiar at any given point of the earth's surface than the terrestrial forms of the same classes, strikingly illustrated in the shells, and in a lesser degree in the insects of this archipelago.

Two of the waders are rather smaller than the same species brought from other places: the swallow is also smaller, though it is doubtful whether or not it is distinct from its analogue. The two owls, the two tyrant flycatchers (*Pyrocephalus*) and the dove, are also smaller than the analogous but distinct species, to which they are most nearly related; on the other hand, the gull is rather larger. The

Figure 3.2 Darwin's finches

1. *Geospiza magnirostris.*   2. *Geospiza fortis.*
3. *Geospiza parvula.*   4. *Certhidea olivacea.*

two owls, the swallow, all three species of mocking-thrush, the dove in its separate colours though not in its whole plumage, the *Totanus*,[2] and the gull, are likewise duskier coloured than their analogous species; and in the case of the mocking-thrush and *Totanus*, than any other species of the two genera. With the exception of a wren with a fine yellow breast, and of a tyrant fly-catcher with a scarlet tuft and breast, none of the birds are brilliantly coloured, as might have been expected in an equatorial district. Hence it would appear probable, that the same causes which here make the immigrants of some species smaller, make most of the peculiar Galapageian species also smaller, as well as very generally more dusky coloured. All the plants have a wretched, weedy appearance, and I did not see one beautiful flower. The insects, again, are small sized and dull coloured, and, as Mr. Waterhouse informs me, there is nothing in their general appearance which would have led him to imagine that they had come from under the equator. The birds, plants, and insects have a desert character, and are not more brilliantly coloured than those from southern Patagonia; we may, therefore, conclude that the usual gaudy colouring of the inter-tropical productions, is not related either to the heat or light of those zones, but to some other cause, perhaps to the conditions of existence being generally favourable to life.

We will now turn to the order of reptiles, which gives the most striking character to the zoology of these islands. The species are not numerous, but the numbers of individuals of each species are extraordinarily great. There is one small lizard belonging to a South American genus, and two species (and probably more) of the *Amblyrhynchus*—a genus confined to the Galapagos islands. There is one snake which is numerous; it is identical, as I am informed by M. Bibron, with the *Psammophis temminckii* from Chile. Of sea-turtle, I believe there is more than one species, and of tortoises there are, as we shall presently show, two or three species or races. Of toads and frogs there are none: I was surprised at this, considering how well suited for them the temperate and damp upper woods appeared to be. It recalled to my mind the remark made by Bory St. Vincent, namely, that none of this family are found on any of the volcanic islands in the great oceans. As far as I can ascertain from various works, this seems to hold good throughout the Pacific, and even in the large islands of the Sandwich archipelago. Mauritius offers an apparent exception, where I saw the *Rana mascariensis* in abundance: this frog is said now to inhabit the Seychelles, Madagascar, and Bourbon; but on the other hand, Du Bois, in his voyage of 1669, states that there were no reptiles in Bourbon except tortoises; and the Officier du Roi asserts that before 1768 it had been attempted, without success, to introduce frogs into Mauritius—I presume, for the purpose of eating: hence it may be well doubted whether this frog is an aboriginal of these islands. The absence of the frog family in the oceanic islands is the more remarkable, when contrasted with the case of lizards, which swarm on most of the smallest islands. May this difference not be caused, by the greater facility with which the eggs of lizards, protected by calcareous shells, might be transported through salt-water, than could the slimy spawn of frogs?

I will first describe the habits of the tortoise (*Testudo nigra*, formerly called *Indica*), which has been so frequently alluded to. These animals are found, I believe, on all the islands of the Archipelago; certainly on the greater number. They frequent in preference the high damp parts, but they likewise live in the lower and arid districts. I have already shown, from the numbers which have been caught in a single day, how very numerous they must be. Some grow to an immense size: Mr. Lawson, an Englishman, and vice-governor of the colony, told us that he had seen several so large, that it required six or eight men to lift them from the ground; and that some had afforded as much as two hundred pounds of meat. The old males are the largest, the females rarely growing to so great a size: the male can be readily distinguished from the female by the greater length of its tail. The tortoises which live on those islands where there is no water, or in the lower and arid parts of the others, feed chiefly on the succulent cactus. Those which frequent the higher and damp regions, eat the leaves of various trees, a kind of berry (called guayavita) which is acid and austere, and likewise a pale green filamentous lichen (*Usnera plicata*), that hangs in tresses from the boughs of the trees.

The tortoise is very fond of water, drinking large quantities, and wallowing in the mud. The larger islands alone possess springs, and these are always situated towards the central parts, and at a considerable height. The tortoises, therefore, which frequent the lower districts, when thirsty, are obliged to travel from a long distance. Hence broad and well-beaten paths branch off in every direction from the wells down to the sea-coast; and the Spaniards by following them up, first discovered the watering-places. When I landed at Chatham Island, I could not imagine what animal travelled so methodically along well-chosen tracks. Near the springs it was a curious spectacle to behold many of these huge creatures, one set eagerly travelling onwards with outstretched necks, and another set returning, after having drunk their fill. When the tortoise arrives at the spring, quite regardless of any spectator, he buries his head in the water above his eyes, and greedily swallows great mouthfuls, at the rate of about ten in a minute. The inhabitants say each animal stays three or four days in the neighbourhood of the water, and then returns to the lower country, but they differed respecting the frequency of these visits. The animal probably regulates them according to the nature of the food on which it has lived. It is, however, certain, that tortoises can subsist even on those islands, where there is no other water than what falls during a few rainy days in the year.

I believe it is well ascertained, that the bladder of the frog acts as a reservoir for the moisture necessary to its existence: such seems to be the case with the tortoise. For some time after a visit to the springs, their urinary bladders are distended with fluid, which is said gradually to decrease in volume, and to become less pure. The inhabitants, when walking in the lower district, and overcome with thirst, often take advantage of this circumstance, and drink the contents of the bladder if full: in one I saw killed, the fluid was quite limpid, and had only a very slightly bitter taste. The inhabitants, however, always first drink the water in the pericardium, which is described as being best.

The tortoises, when purposely moving towards any point, travel by night and day, and arrive at their journey's end much sooner than would be expected. The inhabitants, from observing marked individuals, consider that they travel a distance of about eight miles in two or three days. One large tortoise, which I watched, walked at the rate of sixty yards in ten minutes, that is 360 yards in the hour, or four miles a day,—allowing a little time for it to eat on the road. During the breeding season, when the male and female are together, the male utters a hoarse roar or bellowing, which, it is said, can be heard at the distance of more than a hundred yards. The female never uses her voice, and the male only at these times; so that when the people hear this noise, they know that the two are together. They were at this time (October) laying their eggs. The female, where the soil is sandy, deposits them together, and covers them up with sand; but where the ground is rocky she drops them indiscriminately in any hole: Mr. Bynoe found seven placed in a fissure. The egg is white and spherical; one which I measured was seven inches and three-eighths in circumference, and therefore larger than a hen's egg. The young tortoises, as soon as they are hatched, fall a prey in great numbers to the carrion-feeding buzzard. The old ones seem generally to die from accidents, as from falling down precipices: at least, several of the inhabitants told me, that they had never found one dead without some evident cause.

The inhabitants believe that these animals are absolutely deaf; certainly they do not overhear a person walking close behind them. I was always amused when overtaking one of these great monsters, as it was quietly pacing along, to see how suddenly, the instant I passed, it would draw in its head and legs, and uttering a deep hiss fall to the ground with a heavy sound, as if struck dead. I frequently got on their backs, and then giving a few raps on the hinder part of their shells, they would rise up and walk away;—but I found it very difficult to keep my balance. The flesh of this animal is largely employed, both fresh and salted; and a beautifully clear oil is prepared from the fat. When a tortoise is caught, the man makes a slit in the skin near its tail, so as to see inside its body, whether the fat under the dorsal plate is thick. If it is not, the animal is liberated; and it is said to recover soon from this strange operation. In order to secure the tortoises, it is not sufficient to turn them like turtle, for they are often able to get on their legs again.

There can be little doubt that this tortoise is an aboriginal inhabitant of the Galapagos; for it is found on all, or nearly all, the islands, even on some of the smaller ones where there is no water; had it been an imported species, this would hardly have been the case in a group which has been so little frequented. Moreover, the old Bucaniers found this tortoise in greater numbers even than at present: Wood and Rogers also, in 1708, say that it is the opinion of the Spaniards, that it is found nowhere else in this quarter of the world. It is now widely distributed; but it may be questioned whether it is in any other place an aboriginal. The bones of a tortoise at Mauritius, associated with those of the extinct Dodo, have generally been considered as belonging to this tortoise: if this had been so, undoubtedly it must have been there indigenous; but M. Bibron informs me that he believes that it was distinct, as the species now living there certainly is.

The *Amblyrhynchus*, a remarkable genus of lizards, is confined to this archipelago: there are two species, resembling each other in general form, one being terrestrial and the other aquatic. This latter species (*A. cristatus*) was first characterized by Mr. Bell, who well foresaw, from its short, broad head, and strong claws of equal length, that its habits of life would turn out very peculiar, and different from those of its nearest ally, the Iguana. It is extremely common on all the islands throughout the group, and lives exclusively on the rocky sea-beaches, being never found, at least I never saw one, even ten yards in-shore. It is a hideous-looking creature, of a dirty black colour, stupid, and sluggish in its movements. The usual length of a full-grown one is about a yard, but there are some even four feet long; a large one weighed twenty pounds: on the island of Albemarle they seem to grow to a greater size than elsewhere. Their tails are flattened sideways, and all four feet partially webbed. They are occasionally seen some hundred yards from the shore, swimming about; and Captain Collnett, in his Voyage, says, "They go to sea in herds a-fishing, and sun themselves on the rocks; and may be called alligators in miniature." It must not, however, be supposed that they live on fish. When in the water this lizard swims with perfect ease and quickness, by a serpentine movement of its body and flattened tail—the legs being motionless and closely collapsed on its sides. A seaman on board sank one, with a heavy weight attached to it, thinking thus to kill it directly; but when, an hour afterwards, he drew up the line, it was quite active. Their limbs and strong claws are admirably adapted for crawling over the rugged and fissured masses of lava, which everywhere form the coast. In such situations, a group of six or seven of these hideous reptiles may oftentimes be seen on the black rocks, a few feet above the surf, basking in the sun with outstretched legs.

I opened the stomachs of several, and found them largely distended with minced sea-weed (*Ulvæ*), which grows in thin foliaceous expansions of a bright green or a dull red colour. I do not recollect having observed this sea-weed in any quantity on the tidal rocks; and I have reason to believe it grows at the bottom of the sea, at some little distance from the coast. If such be the case, the object of these animals occasionally going out to sea is explained. The stomach contained nothing but the sea-weed. Mr. Bynoe, however, found a piece of crab in one; but this might have got in accidentally, in the same manner as I have seen a caterpillar, in the midst of some lichen, in the paunch of a tortoise. The intestines were large, as in other herbivorous animals. The nature of this lizard's food, as well as the structure of its tail and feet, and the fact of its having been seen voluntarily swimming out at sea, absolutely prove its aquatic habits; yet there is in this respect one strange anomaly, namely, that when frightened it will not enter the water. Hence it is easy to drive these lizards down to any little point overhanging the sea, where they will soon allow a person to catch hold of their tails than jump into the

Figure 3.3 The Galapagos iguana
Amblyrhynchus cristatus. a, Tooth of natural size, and likewise magnified.

water. They do not seem to have any notion of biting; but when much frightened they squirt a drop of fluid from each nostril. I threw one several times as far as I could, into a deep pool left by the retiring tide; but it invariably returned in a direct line to the spot where I stood. It swam near the bottom, with a very graceful and rapid movement, and occasionally aided itself over the uneven ground with its feet. As soon as it arrived near the edge, but still being under water, it tried to conceal itself in the tufts of sea-weed, or it entered some crevice. As soon as it thought the danger was past, it crawled out on the dry rocks, and shuffled away as quickly as it could. I several times caught this same lizard, by driving it down to a point, and though possessed of such perfect powers of diving and swimming, nothing would induce it to enter the water; and as often as I threw it in, it returned in the manner above described. Perhaps this singular piece of apparent stupidity may be accounted for by the circumstance, that this reptile has no enemy whatever on shore, whereas at sea it must often fall a prey to the numerous sharks. Hence, probably, urged by a fixed and hereditary instinct that the shore is its place of safety, whatever the emergency may be, it there takes refuge.

During our visit (in October), I saw extremely few small individuals of this species, and none I should think under a year old. From this circumstance it seems probable that the breeding season had not then commenced. I asked several of the inhabitants if they knew where it laid its eggs: they said that they knew nothing of its propagation, although well acquainted with the eggs of the land kind—a fact, considering how very common this lizard is, not a little extraordinary.

We will now turn to the terrestrial species (*A. Demarlii*), with a round tail, and toes without webs. This lizard, instead of being found like the other on all the islands, is confined to the central part of the archipelago, namely to Albemarle, James, Barrington, and Indefatigable islands. To the southward, in Charles, Hood, and Chatham islands, and to the northward, in Towers, Bindloes, and Abingdon, I neither saw nor heard of any. It would appear as if it had been created in the centre of the archipelago, and thence had been dispersed only to a certain distance. Some of these lizards inhabit the high and damp parts of the islands, but they are much more numerous in the lower and sterile districts near the coast. I cannot give a more forcible proof of their numbers, than by stating that when we were left at James Island, we could not for some time find a spot free from their burrows on which to pitch our single tent. Like their brothers the sea-kind, they are ugly animals, of a yellowish orange beneath, and of a brownish red colour above: from their low facial angle they have a singularly stupid appearance. They are, perhaps, of a rather less size than the marine species; but several of them weighed between ten and fifteen pounds. In their movements they are lazy and half torpid. When not frightened, they slowly crawl along with their tails and bellies dragging on the ground. They often stop, and doze for a minute or two, with closed eyes and hind legs spread out on the parched soil.

They inhabit burrows, which they sometimes make between fragments of lava, but more generally on level patches of the soft sandstone-like tuff. The holes do not appear to be very deep, and they enter the ground at a small angle; so that when walking over these lizard-warrens, the soil is constantly giving way, much to the annoyance of the tired walker. This animal, when making its burrow, works alternately the opposite sides of its body. One front leg for a short time scratches up the soil, and throws it towards the hind foot, which is well placed so as to heave it beyond the mouth of the hole. That side of the body being tired, the other takes up the task, and so on alternately. I watched one for a long time, till half its body was buried; I then walked up and pulled it by the tail; at this it was greatly astonished, and soon shuffled up to see what was the matter; and then stared me in the face, as much as to say, "What made you pull my tail?"

They feed by day, and do not wander far from their burrows; if frightened, they rush to them with a most awkward gait. Except when running down hill, they cannot move very fast, apparently from the lateral position of their legs. They are not at all timorous: when attentively watching any one, they curl their tails, and, raising themselves on their front legs, nod their heads vertically, with a quick movement, and try to look very fierce: but in reality they are not at all so; if one just stamps on the ground, down go their tails, and off they shuffle as quickly as they can. I have frequently observed small fly-eating lizards, when watching anything, nod their heads in precisely the same manner; but I do not at all know for what purpose. If this *Amblyrhynchus* is held and plagued with a stick, it will bite it very severely; but I caught many by the tail, and they never tried to bite me. If two are placed on the ground and held together, they will fight, and bite each other till blood is drawn.

The individuals, and they are the greater number, which inhabit the lower country, can scarcely taste a drop of water throughout the year; but they consume much of the succulent cactus, the branches of which are occasionally broken off by the wind. I several times threw a piece to two or three of them when together; and it was amusing enough to see them trying to seize and carry it away in their mouths, like so many hungry dogs with a bone. They eat very deliberately, but do not chew their food. The little birds are aware how harmless these creatures are: I have seen one of the thick-billed finches picking at one end of a piece of cactus (which is much relished by all the animals of the lower region), whilst a lizard was eating at the other end; and afterwards the little bird with the utmost indifference hopped on the back of the reptile.

I opened the stomachs of several, and found them full of vegetable fibres and leaves of different trees, especially of an acacia. In the upper region they live chiefly on the acid and astringent berries of the guayavita, under which trees I have seen these lizards and huge tortoises feeding together. To obtain the acacia-leaves they crawl up the low stunted trees; and it is not uncommon to see a pair quietly browsing, whilst seated on a branch several feet above the ground. These lizards, when cooked, yield a white meat, which is liked by those whose stomachs soar above all prejudices. Humboldt has remarked that in intertropical South America, all lizards which inhabit dry regions are esteemed delicacies for the table. The inhabitants state that those which inhabit the upper damp parts drink water, but that the others do not, like the tortoises, travel up for it from the lower sterile country. At the time of our visit, the females had within their bodies numerous large, elongated eggs, which they lay in their burrows: the inhabitants seek them for food.

These two species of *Amblyrhynchus* agree, as I have already stated, in their general structure, and in many of their habits. Neither have that rapid movement, so characteristic of the genera *Lacerta* and *Iguana*. They are both herbivorous, although the kind of vegetation on which they feed is so very different. Mr. Bell has given the name to the genus from the shortness of the snout; indeed, the form of the mouth may almost be compared to that of the tortoise: one is led to suppose that this is an adaptation to their herbivorous appetites. It is very interesting thus to find a well-characterized genus, having its marine and terrestrial species, belonging to so confined a portion of the world. The aquatic species is by far the most remarkable, because it is the only existing lizard which lives on marine vegetable productions. As I at first observed, these islands are not so remarkable for the number of the species of reptiles, as for that of the individuals; when we remember the well-beaten paths made by the thousands of huge tortoises—the many turtles—the great warrens of the terrestrial *Amblyrhynchus*—and the groups of the marine species basking on the coast-rocks of every island—we must admit that there is no other quarter of the world where this Order replaces the herbivorous mammalia in so extraordinary a manner. The geologist on hearing this will probably refer back in his mind to the Secondary epochs, when lizards, some herbivorous, some carnivorous, and of dimensions comparable only with our existing whales, swarmed on the land and in the sea. It is, therefore, worthy of his observation, that this archipelago, instead of possessing a humid climate and rank vegetation, cannot be considered otherwise than extremely arid, and, for an equatorial region, remarkably temperate.

To finish with the zoology: the fifteen kinds of sea-fish which I procured here are all new species; they belong to twelve genera, all widely distributed, with the exception of *Prionotus*, of which the four previously known species live on the eastern side of America. Of land-shells I collected sixteen kinds (and two marked varieties), of which, with the exception of one *Helix* found at Tahiti, all are peculiar to this archipelago: a single fresh-water shell (*Paludina*) is common to Tahiti and Van Diemen's Land. Mr. Cuming, before our voyage, procured here ninety species of sea-shells, and this does not include several species not yet specifically examined, of *Trochus*, *Turbo*, *Monodonta*, and *Nassa*. He has been kind enough to give me the following interesting results: of the ninety shells, no less than forty-seven are unknown elsewhere—a wonderful fact, considering how widely distributed sea-shells generally are. Of the forty-three shells found in other parts of the world, twenty-five inhabit the western coast of America, and of these eight are distinguishable as varieties; the remaining eighteen (including one variety) were found by Mr. Cuming in the Low archipelago, and some of them also at the Philippines. This fact of shells from islands in the central part of the Pacific occurring here, deserves notice, for not one single sea-shell is known to be common to the islands of that ocean and to the west coast of America. The space of open sea running north and south off the west coast, separates two quite distinct conchological provinces; but at the Galapagos Archipelago we have a halting-place, where many new forms have been created, and whither these two great conchological provinces have each sent several colonists. The American province has also sent here representative species; for there is a Galapageian species of *Monoceros*, a genus only found on the west coast of America; and there are Galapageian species of *Fissurella* and *Cancellaria*, genera common on the west coast, but not found (as I am informed by Mr. Cuming) in the central islands of the Pacific. On the other hand, there are Galapageian species of *Oniscia* and *Stylifer*, genera common to the West Indies and to the Chinese and Indian seas, but not found either on the west coast of America or in the central Pacific. I may here add, that after the comparison by Messrs. Cuming and Hinds of about 2000 shells from the eastern and western coasts of America, only one single shell was found in common, namely, the *Purpura patula*, which inhabits the West Indies, the coast of Panama, and the Galapagos. We have, therefore, in this quarter of the world, three great conchological sea-provinces, quite distinct, though surprisingly near each other, being separated by long north and south spaces either of land or of open sea.

I took great pains in collecting the insects, but, excepting Tierra del Fuego, I never saw in this respect so poor a country. Even in the upper and damp region I procured very few, excepting some minute Diptera and Hymenoptera, mostly of common mundane forms. As before remarked, the insects, for a tropical region, are of very small size and dull colours. Of beetles I collected twenty-five species (excluding a *Dermestes* and *Corynetes* imported, wherever a ship touches); of these, two belong to the Harpalidæ, two to the Hydrophilidæ, nine to three families of the Heteromera, and the remaining twelve to as many different families. This circumstance of insects (and I may add plants), where few in number, belonging to many different families, is, I believe, very general. Mr. Waterhouse, who has published an account of the insects of this archipelago, and to whom I am indebted for the above details, informs me that there are several new genera; and that of the genera not new, one or two are American, and the rest of mundane distribution. With the exception of a wood-feeding *Apate*, and of one or probably two water-beetles from the American continent, all the species appear to be new.

The botany of this group is fully as interesting as the zoology. Dr. J. Hooker will soon publish in the "Linnean Transactions" a full account of the Flora, and I am much indebted to him for the following details. Of flowering plants there are, as far as at present is known, 185 species, and 40 cryptogamic species, making together 225; of this number I was fortunate enough to bring home 193. Of the flowering plants, 100 are new species, and are probably confined to this archipelago. Dr. Hooker conceives that, of the plants not so confined, at least 10 species found near the cultivated ground at Charles Island, have been imported. It is, I think, surprising that more American species have not been introduced naturally, considering that the distance is only between 500 and 600 miles from the continent; and that (according to Collnett, p. 58) drift-wood, bamboos, canes, and the nuts of a palm, are often washed on the south-eastern shores. The proportion of 100 flowering plants out of 185 (or 175 excluding the imported weeds) being new, is sufficient, I conceive, to make the Galapagos Archipelago a distinct botanical province; but this Flora is not nearly so peculiar as that of St. Helena, nor, as I am informed by Dr. Hooker, of Juan Fernandez. The peculiarity of the Galapageian Flora is best shown in certain families;—thus there are 21 species of Compositæ, of which 20 are peculiar to this archipelago; these belong to twelve genera, and of these genera no less than ten are confined to the archipelago! Dr. Hooker informs me that the Flora has an undoubted Western American character; nor can he detect in it any affinity with that of the Pacific. If, therefore, we except the eighteen marine, the one fresh-water, and one land-shell, which have apparently come here as colonists from the central islands of the Pacific, and likewise the one distinct Pacific species of the Galapageian group of finches, we see that this archipelago, though standing in the Pacific Ocean, is zoologically part of America.

If this character were owing merely to immigrants from America, there would be little remarkable in it; but we see that a vast majority of all the land animals, and that more than half of the flowering plants, are aboriginal productions. It was most striking to be surrounded by new birds, new reptiles, new shells, new insects, new plants, and yet by innumerable trifling details of structure, and even by the tones of voice and plumage of the birds, to have the temperate plains of Patagonia, or the hot dry deserts of Northern Chile, vividly brought before my eyes. Why, on these small points of land, which within a late geological period must have been covered by the ocean, which are formed of basaltic lava, and therefore differ in geological character from the American continent, and which are placed under a peculiar climate,—why were their aboriginal inhabitants, associated, I may add, in different proportions both in kind and number from those on the continent, and therefore acting on each other in a different manner—why were they created on American types of organization? It is probable that the islands of the Cape de Verd group resemble, in all their physical conditions, far more closely the Galapagos Islands than these latter physically resemble the coast of America; yet the aboriginal inhabitants of the two groups are totally unlike; those of the Cape de Verd Islands bearing the impress of Africa, as the inhabitants of the Galapagos Archipelago are stamped with that of America.

I have not as yet noticed by far the most remarkable feature in the natural history of this archipelago; it is, that the different islands to a considerable extent are inhabited by a different set of beings. My attention was first called to this fact by the Vice-Governor, Mr. Lawson, declaring that the tortoises differed from the different islands, and that he could with certainty tell from which island any one was brought. I did not for some time pay sufficient attention to this statement, and I had already partially mingled together the collections from two of the islands. I never dreamed that islands, about fifty or sixty miles apart, and most of them in sight of each other, formed of precisely the same rocks, placed under a quite similar climate, rising to a nearly equal height, would have been differently tenanted; but we shall soon see that this is the case. It is the fate of most voyagers, no sooner to discover what is most interesting in any locality, than they are hurried from it; but I ought, perhaps, to be thankful that I obtained sufficient material to establish this most remarkable fact in the distribution of organic beings.

The inhabitants, as I have said, state that they can distinguish the tortoises from the different islands; and that they differ not only in size, but in other characters. Captain Porter has described those from Charles and from the nearest island to it, namely, Hood Island, as having their shells in front thick and turned up like a Spanish saddle, whilst the tortoises from James Island are rounder, blacker, and have a better taste when cooked. M. Bibron, moreover, informs me that he has seen what he considers two distinct species of tortoise from the Galapagos, but he does not know from which islands. The specimens that I brought from three islands were young ones; and probably owing to this cause, neither Mr. Gray nor myself could find in them any specific differences. I have remarked that the marine *Amblyrhyn-chus* was larger at Albemarle Island than elsewhere; and M. Bibron informs me that he has seen two distinct aquatic species of this genus; so that the different islands probably have their representative species or races of the *Amblyrhynchus*, as well as of the tortoise. My attention was first thoroughly aroused, by comparing together the numerous specimens, shot by myself and several other parties on board, of the mocking-thrushes, when, to my astonishment, I discovered that all those from Charles Island belonged to one species (*Mimus trifasciatus*); all from Albemarle Island to *M. parvulus*; and all from James and Chatham Islands (between which two other islands are situated, as connecting links) belonged to *M. melanotis*. These two latter species are closely allied, and would by some ornithologists be considered as only well-marked races or varieties; but the *Mimus trifasciatus* is very distinct. Unfortunately most of the specimens of the finch tribe were mingled together; but I have strong reasons to suspect that some of the species of the sub-group *Geospiza* are confined to separate islands. If the different islands have their representatives of *Geospiza*, it may help to explain the singularly large number of the species of this sub-group in this one small archipelago, and as a probable consequence of their numbers, the perfectly graduated series in the size of their beaks. Two species of the sub-group *Cactornis*, and two of *Camarhynchus*, were procured in the archipelago; and of the numerous specimens of these two sub-groups shot by four collectors at James Island, all were found to belong to one species of each; whereas the numerous specimens shot either on Chatham or Charles Island (for the two sets were mingled together) all belonged to the two other species: hence we may feel almost sure that these islands possess their representative species of these two sub-groups. In land-shells this law of distribution does not appear to hold good. In my very small collection of insects,

Mr. Waterhouse remarks, that of those which were ticketed with their locality, not one was common to any two of the islands.

If we now turn to the Flora, we shall find the aboriginal plants of the different islands wonderfully different. I give all the following results on the high authority of my friend Dr. J. Hooker. I may premise that I indiscriminately collected everything in flower on the different islands, and fortunately kept my collections separate. Too much confidence, however, must not be placed in the proportional results, as the small collections brought home by some other naturalists, though in some respects confirming the results, plainly show that much remains to be done in the botany of this group: the Leguminosæ, moreover, have as yet been only approximately worked out:—

Name of Island	Total No. of Species	No. of Species found in other parts of the world	No. of Species confined to the Galapagos Archipelago	No. confined to the one Island	No. of Species confined to the Galapagos Archipelago, but found on more than the one Island
James Island	71	33	38	30	8
Albemarle Island	46	18	26	22	4
Chatham Island	32	16	16	12	4
Charles Island	68	39 (or 29, if the probably imported plants be subtracted)	29	21	8

Hence we have the truly wonderful fact, that in James Island, of the thirty-eight Galapageian plants, or those found in no other part of the world, thirty are exclusively confined to this one island; and in Albemarle Island, of the twenty-six aboriginal Galapageian plants, twenty-two are confined to this one island, that is, only four are at present known to grow in the other islands of the archipelago; and so on, as shown in the above table, with the plants from Chatham and Charles Islands. This fact will, perhaps, be rendered even more striking, by giving a few illustrations:—thus, *Scalesia*, a remarkable arborescent genus of the Compositæ, is confined to the archipelago: it has six species; one from Chatham, one from Albemarle, one from Charles Island, two from James Island, and the sixth from one of the three latter islands, but it is not known from which: not one of these six species grows on any two islands. Again, *Euphorbia*, a mundane or widely distributed genus, has here eight species, of which seven are confined to the archipelago, and not one found on any two islands: *Acalypha* and *Borreria*, both mundane genera, have respectively six and seven species, none of which have the same species on two islands, with the exception of one *Borreria*, which does occur on two islands. The species of the Compositæ are particularly local; and Dr. Hooker has furnished me with several other most striking illustrations of the difference of the species on the different islands. He remarks that this law of distribution holds good both with those genera confined to the archipelago, and those distributed in other quarters of the world: in like manner we have seen that the different islands have their proper species of the mundane genus of tortoise, and of the widely distributed American genus of the mocking-thrush, as well as of two of the Galapageian sub-groups of finches, and almost certainly of the Galapageian genus *Amblyrhynchus*.

The distribution of the tenants of this archipelago would not be nearly so wonderful, if, for instance, one island had a mocking-thrush, and a second island some other quite distinct genus;—if one island had its genus of lizard, and a second island another distinct genus, or none whatever;—or if the different islands were inhabited, not by representative species of the same genera of plants, but by totally different genera, as does to a certain extent hold good; for, to give one instance, a large berry-bearing tree at James Island has no representative species in Charles Island. But it is the circumstance, that several of the islands possess their own species of the tortoise, mocking-thrush, finches, and numerous plants, these species having the same general habits, occupying analogous situations, and obviously filling the same place in the natural economy of this archipelago, that strikes me with wonder. It may be suspected that some of these representative species, at least in the case of the tortoise and of some of the birds, may hereafter prove to be only well-marked races; but this would be of equally great interest to the philosophical naturalist. I have said that most of the islands are in sight of each other: I may specify that Charles Island is fifty miles from the nearest part of Chatham Island, and thirty-three miles from the nearest part of Albemarle Island. Chatham Island is sixty miles from the nearest part of James Island, but there are two intermediate islands between them which were not visited by me. James Island is only ten miles from the nearest part of Albemarle Island, but the two points where the collections were made are thirty-two miles apart. I must repeat, that neither the nature of the soil, nor height of the land, nor the climate, nor the general character of the associated beings, and therefore their action one on another, can differ much in the different islands. If there be any sensible difference in their climates, it must be between the windward group (namely Charles and Chatham Islands), and that to leeward; but there seems to be no corresponding difference in the productions of these two halves of the archipelago.

The only light which I can throw on this remarkable difference in the inhabitants of the different islands, is, that very strong currents of the sea running in a westerly and W.N.W. direction must separate, as far as transportal by the sea is concerned, the southern islands from the northern ones; and between these northern islands a strong N.W. current was observed, which must effectually separate James and Albemarle Islands. As the archipelago is free to a most remarkable degree from gales of wind, neither the birds, insects, nor lighter seeds, would be blown from island to island. And lastly, the profound depth of the ocean between the islands, and their apparently recent (in a geological sense) volcanic origin, render it highly unlikely that they were ever united; and this, probably, is a far more important consideration than any other, with respect to the geographical distribution of their inhabitants. Reviewing the facts here given, one is astonished at the amount of creative force, if such an expression may be used, displayed on these small, barren, and rocky islands; and still more so, at its diverse yet analogous action on points so near each other. I have said that the Galapagos Archipelago might be called a satellite attached to America, but it should rather be called a group of satellites, physically similar, organically distinct, yet intimately related to each other, and all related in a marked, though much lesser degree, to the great American continent.

I will conclude my description of the natural history of these islands, by giving an account of the extreme tameness of the birds.

This disposition is common to all the terrestrial species; namely, to the mocking-thrushes, the finches, wrens, tyrant-flycatchers, the dove, and carrion-buzzard. All of them often approached sufficiently near to be killed with a switch, and sometimes, as I myself tried, with a cap or hat. A gun is here almost superfluous; for with the muzzle I pushed a hawk off the branch of a tree. One day, whilst lying down, a mocking-thrush alighted on the edge of a pitcher, made of the shell of a tortoise, which I held in my hand, and began very quietly to sip the water; it allowed me to lift it from the ground whilst seated on the vessel: I often tried, and very nearly succeeded, in catching these birds by their legs. Formerly the birds appear to have been even tamer than at present. Cowley (in the year 1684) says that the 'Turtle-doves were so tame, that they would often alight upon our hats and arms, so as that we could take them alive: they not fearing man, until such time as some of our company did fire at them, whereby they were rendered more shy." Dampier also, in the same year, says that a man in a morning's walk might kill six or seven dozen of these doves. At present, although certainly very tame, they do not alight on people's arms, nor do they suffer themselves to be killed in such large numbers. It is surprising that they have not become wilder; for these islands during the last hundred and fifty years have been frequently visited by bucaniers and whalers; and the sailors, wandering through the woods in search of tortoises, always take cruel delight in knocking down the little birds.

These birds, although now still more persecuted, do not readily become wild: in Charles Island, which had then been colonized about six years, I saw a boy sitting by a well with a switch in his hand, with which he killed the doves and finches as they came to drink. He had already procured a little heap of them for his dinner; and he said that he had constantly been in the habit of waiting by this well for the same purpose. It would appear that the birds of this archipelago, not having as yet learnt that man is a more dangerous animal than the tortoise or the *Amblyrhynchus*, disregard him, in the same manner as in England shy birds, such as magpies, disregard the cows and horses grazing in our fields.

The Falkland Islands offer a second instance of birds with a similar disposition. The extraordinary tameness of the little *Opetiorhynchus* has been remarked by Pernety, Lesson, and other voyagers. It is not, however, peculiar to that bird: the *Polyborus*, snipe, upland and lowland goose, thrush, bunting, and even some true hawks, are all more or less tame. As the birds are so tame there, where foxes, hawks, and owls occur, we may infer that the absence of all rapacious animals at the Galapagos, is not the cause of their tameness here. The upland geese at the Falklands show, by the precaution they take in building on the islets, that they are aware of their danger from the foxes; but they are not by this rendered wild towards man. This tameness of the birds, especially of the waterfowl, is strongly contrasted with the habits of the same species in Tierra del Fuego, where for ages past they have been persecuted by the wild inhabitants. In the Falklands, the sportsman may sometimes kill more of the upland geese in one day than he can carry home; whereas in Tierra del Fuego, it is nearly as difficult to kill one, as it is in England to shoot the common wild goose.

In the time of Pernety (1763), all the birds there appear to have been much tamer than at present; he states that the *Opetiorhynchus* would almost perch on his finger; and that with a wand he killed ten in half an hour. At that period the birds must have been about as tame, as they now are at the Galapagos. They appear to have learnt caution more slowly at these latter islands than at the Falklands, where they have had proportionate means of experience; for besides frequent visits from vessels, those islands have been at intervals colonized during the entire period. Even formerly, when all the birds were so tame, it was impossible by Pernety's account to kill the black-necked swan—a bird of passage, which probably brought with it the wisdom learnt in foreign countries.

I may add that, according to Du Bois, all the birds at Bourbon in 1571–72, with the exception of the flamingoes and geese, were so extremely tame, that they could be caught by the hand, or killed in any number with a stick. Again, at Tristan d'Acunha in the Atlantic, Carmichael[3] states that the only two land-birds, a thrush and a bunting, were "so tame as to suffer themselves to be caught with a hand-net." From these several facts we may, I think, conclude, first, that the wildness of birds with regard to man, is a particular instinct directed against *him*, and not dependent on any general degree of caution arising from other sources of danger; secondly, that it is not acquired by individual birds in a short time, even when much persecuted; but that in the course of successive generations it becomes hereditary. With domesticated animals we are accustomed to see new mental habits or instincts acquired and rendered hereditary; but with animals in a state of nature, it must always be most difficult to discover instances of acquired hereditary knowledge. In regard to the wildness of birds towards man, there is no way of accounting for it, except as an inherited habit: comparatively few young birds, in any one year, have been injured by man in England, yet almost all, even nestlings, are afraid of him; many individuals, on the other hand, both at the Galapagos and at the Falklands, have been pursued and injured by man, but yet have not learned a salutary dread of him. We may infer from these facts, what havoc the introduction of any new beast of prey must cause in a country, before the instincts of the indigenous inhabitants have become adapted to the stranger's craft or power.

Notes

1 Which is positioned in the Atlantic Ocean, off the coast of Brazil just below the Equator. It is therefore geographically the Atlantic mirror-image of the Galapagos in the Pacific.

2 *Totanus* is a genus of sandpipers. Two species are occasionally seen there, but neither breed on the islands. No waders breed there. The other birds mentioned here by Darwin do all breed on the Galapagos.

3 [Darwin's note] Linn. Trans., vol. xii. p. 496. The most anomalous fact on this subject which I have met with, is the wildness of the small birds in the Arctic parts of North America (as described by Richardson, Fauna Bor., vol. ii. p. 332), where they are said never to be persecuted. This case is the more strange, because it is asserted that some of the same species in their winter-quarters in the United States are tame. There is much, as Dr. Richardson well remarks, utterly inexplicable connected with the different degrees of shyness and care with which birds conceal their nests. How strange it is that the English wood-pigeon, generally so wild a bird, should very frequently rear its young in shrubberies close to houses!

———————— *Chapter four*————————

The origin of species
(1859)

If everything had gone according to plan, the *Origin* would never have been written. Darwin had been developing his theory of evolution since 1837; he was in no hurry, as he did not fear pre-emption. The subject, he believed, was not 'in the air': 'I occasionally sounded not a few naturalists, and never came across a single one who seemed to doubt the permanence of species.' But in the summer of 1858 he was forced to change gear. Here (from his *Autobiography*, with some sentences omitted) is how he later told the story of one of the most dramatic cases of independent discovery in the history of science.

Early in 1856 Lyell advised me to write out my views pretty fully, and I began at once to do so on a scale three or four times as extensive as that which was afterwards followed in my Origin of Species; yet it was only an abstract of the materials I had collected, and I got through about half the work on this scale. But my plans were overthrown, for early in the summer of 1858 Mr Wallace, who was then in the Malay Archipelago, sent me an essay 'On the tendency of varieties to depart indefinitely from the original type'; and this essay contained exactly the same theory as mine. Mr Wallace expressed the wish that if I thought well of his essay, I should send it to Lyell for perusal. The circumstances under which I consented at the request of Lyell and Hooker to allow an extract from my M.S, together with a letter to Asa Gray dated September 5 1857, to be published at the same time with Wallace's Essay, are given in the Journal of the Proceedings of the Linn. Soc. 1858 p. 45. I was at first very unwilling to consent, as I thought Mr. Wallace might consider my doing so unjustifiable, for I did not then know how generous and noble was his disposition. The extract from my M.S and the letter to Asa Gray had neither been intended for publication and were badly written. Mr. Wallace's essay, on the other hand was admirably expressed and quite clear. Nevertheless our joint productions excited very little attention, and the only published notice of them which I can remember was by Prof. Haughton of Dublin, whose verdict was that all that was new in them was false, and what was true was old. This shows how necessary it is that any new view should be explained at considerable length in order to arouse public attention.

In September 1858 I set to work by the strong advice of Lyell and Hooker to prepare a volume on the transmutation of species, but was often interrupted by ill-health ... It cost me 13 months and ten days hard labour. It was published under the title of the 'Origin of Species' in November 1859 ... It is no doubt the chief work of my life.

In its first edition, the *Origin* is 14 chapters long. It begins with three chapters on inheritance, two that expound the theory of natural selection, and one that considers the main difficulties of the theory; a chapter on 'instinct' then applies the theory to behaviour. In Chapters 8–14, which might be thought of as making up the second half of the book, Darwin critically considers the evidence for evolution rather than the separate creation of species. He discusses evidence from the fossil record, geographical distribution, comparative anatomy and embryology. 'The facts seem to me to come out *very* strong for mutability of species', he wrote to Hooker in January 1859. Although a modern account of natural selection would be more genetical in tone than Darwin's original account, his theory remains effectively unchallenged by more than a century of work, and his review of the case for evolution is still the best and most comprehensive ever written. The *Origin* made so strong a case for evolution that most biologists were soon persuaded, and have not subsequently felt the need to write a modern version, in which recent discoveries could be included. Biologists took rather longer to accept Darwin's theory of natural selection.

I have extracted much of Darwin's beautifully clear exposition of the theory of natural selection (from Chapters 3 and 4 of the *Origin*). They contain some of his best writing (as, for example, the passages on the tangled bank, p. 81, and the relation of cats and clover, p. 81) and are thoroughly modern in conception. The second of these chapters includes the section on the principle of divergence, the significance of which I discussed in the introductory chapter (p. 14). The argument for evolution is longer, as separate chapters consider the evidence of biogeography, the geological succession, classification, and morphology. One chapter counters the apparently contrary evidence of the geological record; Darwin argues that the record is too incomplete to demonstrate that species have not changed through time. The final chapter of the *Origin* recapitulates the case for evolution, and I have extracted the relevant passage here. The only difficulty is that Darwin supposes the reader of the recapitulation to have read the longer version of his preceding six chapters, and therefore discusses many of the facts only allusively or synoptically. But I think it is possible to follow the outline of the argument, and have annotated the obscure allusions.

So there follow two passages on natural selection, and an abstract of the case for evolution. When Darwin began the book, he described it to Hooker as 'amusing and improving work'. And then, in early 1859 he told his publisher, 'It may be conceit, but I believe the subject will interest the public.'

Struggle for existence

Before entering on the subject of this chapter, I must make a few preliminary remarks, to show how the struggle for existence bears on Natural Selection.... Amongst organic beings in a state of nature there is some individual variability;[1] indeed I am not aware that this has ever been disputed. It is immaterial for us whether a multitude of doubtful forms be called species or sub-species or varieties; what rank, for instance, the two or three hundred doubtful forms of British plants are entitled to hold, if the existence of any well-marked varieties be admitted. But the mere existence of individual variability and of some few well-marked varieties, though necessary as the foundation for the work, helps us but little in understanding how species arise in nature. How have all those exquisite adaptations of one part of the organisation to another part, and to the conditions of life, and of one distinct organic being to another being, been perfected? We see these beautiful co-adaptations most plainly in the woodpecker and mistletoe; and only a little less plainly in the humblest parasite which clings to the hairs of a quadruped or feathers of a bird; in the structure of the beetle which dives through the water; in the plumed seed which is wafted by the gentlest breeze; in short, we see beautiful adaptations everywhere and in every part of the organic world.

Again, it may be asked, how is it that varieties, which I have called incipient species, become ultimately converted into good and distinct species, which in most cases obviously differ from each other far more than do the varieties of the same species? How do those groups of species, which constitute what are called distinct genera, and which differ from each other more than do the species of the same genus, arise? All these results, as we shall more fully see in the next chapter, follow inevitably from the struggle for life. Owing to this struggle for life, any variation, however slight and from whatever cause proceeding, if it be in any degree profitable to an individual of any species, in its infinitely complex relations to other organic beings and to external nature, will tend to the preservation of that individual, and will generally be inherited by its offspring. The offspring, also, will thus have a better chance of surviving, for, of the many individuals of any species which are periodically born, but a small number can survive. I have called this principle, by which each slight variation, if useful, is preserved, by the term of Natural Selection, in order to mark its relation to man's power of selection. . . . Man by selection can certainly produce great results, and can adapt organic beings to his own uses, through the accumulation of slight but useful variations, given to him by the hand of Nature. But Natural Selection, as we shall hereafter see, is a power incessantly ready for action, and is as immeasurably superior to man's feeble efforts, as the works of Nature are to those of Art.

We will now discuss in a little more detail the struggle for existence. In my future work this subject shall be treated, as it well deserves, at much greater length. The elder De Candolle[2] and Lyell have largely and philosophically shown that all organic beings are exposed to severe competition. In regard to plants, no one has treated this subject with more spirit and ability than W. Herbert, Dean of Manchester, evidently the result of his great horticultural knowledge. Nothing is easier than to admit in words the truth of the universal struggle for life, or more difficult—at least I have found it so—than constantly to bear this conclusion in mind. Yet unless it be thoroughly engrained in the mind, I am convinced that the whole economy of nature, with every fact on distribution, rarity, abundance, extinction, and variation, will be dimly seen or quite misunderstood. We behold the face of nature bright with gladness, we often see superabundance of food; we do not see, or we forget, that the birds which are idly singing round us mostly live on insects or seeds, and are thus constantly destroying life; or we forget how largely these songsters, or their eggs, or their nestlings, are destroyed by birds and beasts of prey; we do not always bear in mind, that though food may be now superabundant, it is not so at all seasons of each recurring year.

I should premise that I use the term Struggle for Existence in a large and metaphorical sense, including dependence of one being on another, and including (which is more important) not only the life of the individual, but success in leaving progeny. Two canine animals in a time of dearth, may be truly said to struggle with each other which shall get food and live. But a plant on the edge of a desert is said to struggle for life against the drought, though more properly it should be said to be dependent on the moisture. A plant which annually produces a thousand seeds, of which on an average only one comes to maturity, may be more truly said to struggle with the plants of the same and other kinds which already clothe the ground. The mistletoe is dependent on the apple and a few other trees, but can only in a far-fetched sense be said to struggle with these trees, for if too many of these parasites grow on the same tree, it will languish and die. But several seedling mistletoes, growing close together on the same branch, may more truly be said to struggle with each other. As the mistletoe is disseminated by birds, its existence depends on birds; and it may metaphorically be said to struggle with other fruit-bearing plants, in order to tempt birds to devour and thus disseminate its seeds rather than those of other plants. In these several senses, which pass into each other, I use for convenience sake the general term of struggle for existence.

A struggle for existence inevitably follows from the high rate at which all organic beings tend to increase. Every being, which during its natural lifetime produces several eggs or seeds, must suffer destruction during some period of its life, and during some season or occasional year, otherwise, on the principle of geometrical increase, its numbers would quickly become so inordinately great that no country could support the product. Hence, as more individuals are produced than can possibly survive, there must in every case be a struggle for existence, either one individual with another of the same species, or with the individuals of distinct species, or with the physical conditions of life. It is the doctrine of Malthus applied with manifold force to the whole animal and vegetable kingdoms; for in this case there can be no artificial increase of food, and no prudential restraint from marriage. Although some species may be now increasing, more or less rapidly, in numbers, all cannot do so, for the world would not hold them.

There is no exception to the rule that every organic being naturally increases at so high a rate, that if not destroyed, the earth would soon be covered by the progeny of a single pair. Even slow-breeding man has doubled in twenty-five years, and at this rate, in a few thousand years, there would literally not be standing room for his progeny. Linnaeus has calculated that if an annual plant produced only two seeds—and there is no plant so unproductive as this—and their seedlings next year produced two, and so on, then in twenty years there would be a million plants. The elephant is reckoned to be the slowest breeder of all known animals, and I have taken some pains to estimate its probable minimum rate of natural increase: it will be under the mark to assume that it breeds when thirty years old, and goes on breeding till ninety years old, bringing forth three pairs of young in this interval; if this be so, at the end of the fifth century there would be alive fifteen million elephants, descended from the first pair.[3]

But we have better evidence on this subject than mere theoretical calculations, namely, the numerous recorded cases of the astonishingly rapid increase of various animals in a state of nature, when circumstances have been favourable to them during two or three following seasons. Still more striking is the evidence from our domestic animals of many kinds which have run wild in several parts of the world: if the statements of the rate of increase of slow-breeding cattle and horses in South-America, and latterly in Australia, had not been well authenticated, they would have been quite incredible. So it is with plants: cases could be given of introduced plants which have become common throughout whole islands in a period of less than ten years. Several of the plants now most numerous over the wide plains of La Plata, clothing square leagues of surface almost to the exclusion of all other plants, have been introduced from Europe; and there are plants which now range in India, as I hear from Dr. Falconer, from Cape Comorin to the Himalaya, which have been imported from America since its discovery. In such cases, and endless instances could be given, no one supposes that the fertility of these animals or plants has been suddenly and temporarily increased in any sensible degree. The obvious explanation is that the conditions of life have been very favourable, and that there has consequently been less destruction of the old and young, and that nearly all the young have been enabled to breed. In such cases the geometrical ratio of increase, the result of which never fails to be surprising, simply explains the extraordinarily rapid increase and wide diffusion of naturalised productions in their new homes.

In a state of nature almost every plant produces seed, and amongst animals there are very few which do not annually pair. Hence we may confidently assert, that all plants and animals are tending to increase at a geometrical ratio, that all would most rapidly stock every station in which they could any how exist, and that the geometrical tendency to increase must be checked by destruction at some period of life. Our familiarity with the larger domestic animals tends, I think, to mislead us: we see no great destruction falling on them, and we forget that thousands are annually slaughtered for food, and that in a state of nature an equal number would have somehow to be disposed of.

The only difference between organisms which annually produce eggs or seeds by the thousand, and those which produce extremely few, is, that the slow-breeders would require a few more years to people, under favourable conditions, a whole district, let it be ever so large. The condor lays a couple of eggs and the ostrich a score, and yet in the same country the condor may be the more numerous of the two: the Fulmar petrel lays but one egg, yet it is believed to be the most numerous bird in the world. One fly deposits hundreds of eggs, and another, like the *Hippobosca*, a single one; but this difference does not determine how many individuals of the two species can be supported in a district. A large number of eggs is of some importance to those species, which depend on a rapidly fluctuating amount of food, for it allows them rapidly to increase in number. But the real importance of a large number of eggs or seeds is to make up for much destruction at some period of life; and this period in the great majority of cases is an early one. If an animal can in any way protect its own eggs or young, a small number may be produced, and yet the average stock be fully kept up; but if many eggs or young are destroyed, many must be produced, or the species will become extinct. It would suffice to keep up the full number of a tree, which lived on an average for a thousand years, if a single seed were produced once in a thousand years, supposing that this seed were never destroyed, and could be ensured to germinate in a fitting place. So that in all cases, the average number of any animal or plant depends only indirectly on the number of its eggs or seeds.

In looking at Nature, it is most necessary to keep the foregoing considerations always in mind—never to forget that every single organic being around us may be said to be striving to the utmost to increase in numbers; that each lives by a struggle at some period of its life; that heavy destruction inevitably falls either on the young or old, during each generation or at recurrent intervals. Lighten any check, mitigate the destruction ever so little, and the number of the species will almost instantaneously increase to any amount. The face of Nature may be compared to a yielding surface, with ten thousand sharp wedges packed close together and driven inwards by incessant blows, sometimes one wedge being struck, and then another with greater force.

What checks the natural tendency of each species to increase in number is most obscure. Look at the most vigorous species; by as much as it swarms in numbers, by so much will its tendency to increase be still further increased. We know not exactly what the checks are in even one single instance. Nor will this surprise any one who reflects how ignorant we are on this head, even in regard to mankind, so incomparably better known than any other animal. . . . Here I will make only a few remarks, just to recall to the reader's mind some of the chief points. Eggs or very young animals seem generally to suffer most, but this is not invariably the case. With plants there is a vast destruction of seeds, but, from some observations which I have made, I believe that it is the seedlings which suffer most from germinating in ground already thickly stocked with other plants. Seedlings, also, are destroyed in vast numbers by various enemies; for instance, on a piece of ground three feet long and two wide, dug and cleared, and where there could be no choking from other plants, I marked all the seedlings of our native weeds as they came up, and out of the 357 no less than 295 were destroyed, chiefly by slugs and insects. If turf which has long been mown, and the case would be the same with turf closely browsed by quadrupeds, be let to grow, the more vigorous plants gradually kill the less vigorous, though fully grown, plants: thus out of twenty species growing on a little plot of turf (three feet by four) nine species perished from the other species being allowed to grow up freely.

The amount of food for each species of course gives the extreme limit to which each can increase; but very frequently it is not the obtaining food, but the serving as prey to other animals, which determines the average numbers of a species. Thus, there seems to be little doubt that the stock of partridges, grouse, and hares on any large estate depends chiefly on the destruction of vermin. If not one head of game were shot during the next twenty years in England, and, at the same time, if no vermin were destroyed, there would, in all probability, be less game than at present, although hundreds of thousands of game animals are now annually killed. On the other hand, in some cases, as with the elephant and rhinoceros, none are destroyed by beasts of prey: even the tiger in India most rarely dares to attack a young elephant protected by its dam.

Climate plays an important part in determining the average numbers of a species, and periodical seasons of extreme cold or drought, I believe to be the most effective of all checks. I estimated that the winter of 1854–55 destroyed four-fifths of the birds in my own grounds; and this is a tremendous destruction, when we remember that ten per cent. is an extraordinarily severe mortality from epidemics with man. The action of climate seems at first sight to be quite independent of the struggle for existence; but in so far as climate chiefly acts in reducing food, it brings on the most severe struggle between the individuals, whether of the same or of distinct species, which subsist on the same kind of food. Even when climate, for instance extreme cold, acts directly, it will be the least vigorous, or those which have got least food through the advancing winter, which will suffer most. When we travel from south to north, or from a damp region to a dry, we invariably see some species gradually getting rarer and rarer, and finally disappearing; and the change of climate being conspicuous, we are tempted to attribute the whole effect to its direct action. But this is a very false view: we forget that each species, even where it most abounds, is constantly suffering enormous destruction at some period of its life, from enemies or from competitors for the same place and food; and if these enemies or competitors be in the least degree favoured by any slight change of climate, they will increase in numbers, and, as each area is already fully stocked with inhabitants, the other species will decrease. When we travel southward and see a species decreasing in numbers, we may feel sure that the cause lies quite as much in other species being favoured, as in this one being hurt. So it is when we travel northward, but in a somewhat lesser degree, for the number of species of all kinds, and therefore of competitors, decreases northwards; hence in going northward, or in ascending a mountain, we far oftener meet with stunted forms, due to the *directly* injurious action of climate, than we do in proceeding southwards or in descending a mountain. When we reach the Arctic regions, or snow-capped summits, or absolute deserts, the struggle for life is almost exclusively with the elements.

That climate acts in main part indirectly by favouring other species, we may clearly see in the prodigious number of plants in our gardens which can perfectly well endure our climate, but which never become naturalised, for they cannot compete with our native plants, nor resist destruction by our native animals.

When a species, owing to highly favourable circumstances, increases inordinately in numbers in a small tract, epidemics—at least, this seems generally to occur with our game animals—often ensue: and here we have a limiting check independent of the struggle for life. But even some of these so-called epidemics appear to be due to parasitic worms, which have from some cause, possibly in part through facility of diffusion amongst the crowded animals, been disproportionably favoured: and here comes in a sort of struggle between the parasite and its prey.

On the other hand, in many cases, a large stock of individuals of the same species, relatively to the numbers of its enemies, is absolutely necessary for its preservation. Thus we can easily raise plenty of corn and rape-seed, &c., in our fields, because the seeds are in great excess compared with the number of birds which feed on them; nor can the birds, though having a superabundance of food at this one season, increase in number proportionally to the supply of seed, as their numbers are checked during winter: but any one who has tried, knows how troublesome it is to get seed from a few wheat or other such plants in a garden; I have in this case lost every single seed. This view of the necessity of a large stock of the same species for its preservation, explains, I believe, some singular facts in nature, such as that of very rare plants being sometimes extremely abundant in the few spots where they do occur; and that of some social plants being social, that is, abounding in individuals, even on the extreme confines of their range. For in such cases, we may believe, that a plant could exist only where the conditions of its life were so favourable that many could exist together, and thus save each other from utter destruction. I should add that the good effects of frequent inter-crossing, and the ill effects of close interbreeding, probably came into play in some of these cases; but on this intricate subject I will not here enlarge.

Many cases are on record showing how complex and unexpected are the checks and relations between organic beings, which have to struggle together in the same country. I will give only a single instance, which, though a simple one, has interested me. In Stafford-shire, on the estate of a relation where I had ample means of investigation, there was a large and extremely barren heath, which had never been touched by the hand of man; but several hundred acres of exactly the same nature had been enclosed twenty-five years previously and planted with Scotch fir. The change in the native vegetation of the planted part of the heath was most remarkable, more than is generally seen in passing from one quite different soil to another: not only the proportional numbers of the heath-plants were wholly changed, but twelve species of plants (not counting grasses and carices) flourished in the plantations, which could not be found on the heath. The effect on the insects must have been still greater, for six insectivorous birds were very common in the plantations, which were not to be seen on the heath; and the heath was frequented by two or three distinct insectivorous birds. Here we see how potent has been the effect of the introduction of a single tree, nothing whatever else having been done, with the exception that the land had been enclosed, so that cattle could not enter. But how important an element enclosure is, I plainly saw near Farnham, in Surrey. Here there are extensive heaths, with a few clumps of old Scotch firs on the distant hill-tops: within the last ten years large spaces have been enclosed, and self-sown firs are now springing up in multitudes, so close together that all cannot live. When I ascertained that these young trees had not been sown or planted, I was so much surprised at their numbers that I went to several points of view, whence I could examine hundreds of acres of the unenclosed heath, and literally I could not see a single Scotch fir, except the old planted clumps. But on looking closely between the stems of the heath, I found a multitude of seedlings and little trees, which had been perpetually browsed down by the cattle. In one square yard, at a point some hundreds yards distant from one of the old clumps, I counted thirty-two little trees; and one of them, judging from the rings of growth, had during twenty-six years tried to raise its head above the stems of the heath, and had failed. No wonder that, as soon as the land was enclosed, it became thickly clothed with vigorously growing young firs. Yet the heath was so extremely barren and so extensive that no one would ever have imagined that cattle would have so closely and effectually searched it for food.

Here we see that cattle absolutely determine the existence of the Scotch fir; but in several parts of the world insects determine the existence of cattle. Perhaps Paraguay offers the most curious instance of this; for here neither cattle nor horses nor dogs have ever run wild, though they swarm southward and northward in a feral state; and Azara and Rengger have shown that this is caused by the greater number in Paraguay of a certain fly, which lays its eggs in the navels of these animals when first born. The increase of these flies, numerous as they are, must be habitually checked by some means, probably by birds. Hence, if certain insectivorous birds (whose numbers are probably regulated by hawks or beasts of prey) were to increase in Paraguay, the flies would decrease—then cattle and horses would become feral, and this would certainly greatly alter (as indeed I have observed in parts of South America) the vegetation: this again would largely affect the insects; and this, as we just have seen in Stafford-shire, the insectivorous birds, and so onwards in ever-increasing circles of complexity. We began this series by insectivorous birds, and we have ended with them. Not that in nature the relations can ever be as simple as this. Battle within battle must ever be recurring with varying success; and yet in the long-run the forces are so nicely balanced, that the face of nature remains uniform for long periods of time, though assuredly the merest trifle would often give the victory to one organic being over another. Nevertheless so profound is our ignorance, and so high our presumption, that we marvel when we hear of the extinction of an organic being; and as we do not see the cause, we invoke cataclysms to desolate the world, or invent laws on the duration of the forms of life!

I am tempted to give one more instance showing how plants and animals, most remote in the scale of nature, are bound together by a web of complex relations. I shall hereafter have occasion to show that the exotic *Lobelia fulgens*, in this part of England, is never visited by insects, and consequently, from its peculiar structure, never can set a seed. Many of our orchidaceous plants absolutely require the visits of moths to remove their pollen-masses and thus to fertilise them. I have, also, reason to believe that humble-bees are indispensable to the fertilisation of the heartsease (*Viola tricolor*), for other bees do not visit this flower. From experiments which I have tried, I have found that the visits of bees, if not indispensable, are at least highly beneficial to the fertilisation of our clovers; but humble-bees alone visit the common red clover (*Trifolium pratense*), as other bees cannot reach the nectar. Hence I have very little doubt, that if the whole genus of humble-bees became extinct or very rare in England, the heartsease and red clover would become very rare, or wholly disappear. The number of humble-bees in any district depends in a great degree on the number of field-mice, which destroy their combs and nests; and Mr. H. Newman, who has long attended to the habits of humble-bees, believes that 'more than two thirds of them are thus destroyed all over England.' Now the number of mice is largely dependent, as every one knows, on the number of cats; and Mr. Newman says, 'Near villages and small towns I have found the nests of humble-bees more numerous than elsewhere, which I attribute to the number of cats that destroy the mice.' Hence it is quite credible that the presence of a feline animal in large numbers in a district might determine, through the intervention first of mice and then of bees, the frequency of certain flowers in that district!

In the case of every species, many different checks, acting at different periods of life, and during different seasons or years, probably come into play; some one check or some few being generally the most potent, but all concurring in determining the average number or even the existence of the species. In some cases it can be shown that widely-different checks act on the same species in different districts. When we look at the plants and bushes clothing an entangled bank, we are tempted to attribute their proportional numbers and kinds to what we call chance. But how false a view is this! Every one has heard that when an American forest is cut down, a very different vegetation springs up; but it has been observed that the trees now growing on the ancient Indian mounds, in the Southern United States, display the same beautiful diversity and proportion of kinds as in the surrounding virgin forests. What a struggle between the several kinds of trees must here have gone on during long centuries, each annually scattering its seeds by the thousand; what war between insect and insect—between insects, snails, and other animals with birds and beasts of prey—all striving to increase, and all feeding on each other or on the trees or their seeds and seedlings, or on the other plants which first clothed the ground and thus checked the growth of the trees! Throw up a handful of feathers, and all must fall to the ground according to definite laws; but how simple is this problem compared to the action and reaction of the innumerable plants and animals which have determined, in the course of centuries, the proportional numbers and kinds of trees now growing on the old Indian ruins!

The dependency of one organic being on another, as of a parasite on its prey, lies generally between beings remote in the scale of nature. This is often the case with those which may strictly be said to struggle with each other for existence, as in the case of locusts and grass-feeding quadrupeds. But the struggle almost invariably will be most severe between the individuals of the same species, for they frequent the same districts, require the same food, and are exposed to the same dangers. In the case of varieties of the same species, the struggle will generally be almost equally severe, and we sometimes see the contest soon decided: for instance, if several varieties of wheat be sown together, and the mixed seed be resown, some of the varieties which best suit the soil or climate, or are naturally the most fertile, will beat the others and so yield more seed, and will consequently in a few years quite supplant the other varieties. To keep up a mixed stock of even such extremely close varieties as the variously coloured sweet-peas, they must be each year harvested separately, and the seed then mixed in due proportion, otherwise the weaker kinds will steadily decrease in numbers and disappear. So again with the varieties of sheep: it has been asserted that certain mountain-varieties will starve out other mountain-varieties, so that they cannot be kept together. The same result has followed from keeping together different varieties of the medicinal leech. It may even be doubted whether the varieties of any one of our domestic plants or animals have so exactly the same strength, habits, and constitution, that the original proportions of a mixed stock could be kept up for half a dozen generations, if they were allowed to struggle together, like beings in a state of nature, and if the seed or young were not annually sorted.

As species of the same genus have usually, though by no means invariably, some similarity in habits and constitution, and always in structure, the struggle will generally be more severe between species of the same genus, when they come into competition with each other, than between species of distinct genera. We see this in the recent extension over parts of the United States of one species of swallow having caused the decrease of another species. The recent increase of the missel-thrush in parts of Scotland has caused the decrease of the song-thrush. How frequently we hear of one species of rat taking the place of another species under the most different climates! In Russia the small Asiatic cockroach has everywhere driven before it its great congener. One species of charlock will supplant another, and so in other cases. We can dimly see why the competition should be most severe between allied forms, which fill nearly the same place in the economy of nature; but probably in no one case could we precisely say why one species has been victorious over another in the great battle of life.

A corollary of the highest importance may be deduced from the foregoing remarks, namely, that the structure of every organic being is related, in the most essential yet often hidden manner, to that of all other organic beings, with which it comes into competition for food or residence, or from which it has to escape, or on which it preys. This is obvious in the structure of the teeth and talons of the tiger; and in that of the legs and claws of the parasite which clings to the hair on the tiger's body. But in the beautifully plumed seed of the dandelion, and in the flattened and fringed legs of the water-beetle, the relation seems at first confined to the elements of air and water. Yet the advantage of plumed seeds no doubt stands in the closest relation to the land being already thickly clothed by other plants; so that the seeds may be widely distributed and fall on unoccupied ground. In the water-beetle, the structure of its legs, so well adapted for diving, allows it to compete with other aquatic insects, to hunt for its own prey, and to escape serving as prey to other animals.

The store of nutriment laid up within the seeds of many plants seems at first sight to have no sort of relation to other plants. But from the strong growth of young plants produced from such seeds (as peas and beans), when sown in the midst of long grass, I suspect that the chief use of the nutriment in the seed is to favour the growth of the young seedling, whilst struggling with other plants growing vigorously all around.

Look at a plant in the midst of its range, why does it not double or quadruple its numbers? We know that it can perfectly well withstand a little more heat or cold, dampness or dryness, for elsewhere it ranges into slightly hotter or colder, damper or drier districts. In this case we can clearly see that if we wished in imagination to give the plant the power of increasing in number, we should have to give it some advantage over its competitors, or over the animals which preyed on it. On the confines of its geographical range, a change of constitution with respect to climate would clearly be an advantage to our plant; but we have reason to believe that only a few plants or

83

animals range so far, that they are destroyed by the rigour of the climate alone. Not until we reach the extreme confines of life, in the arctic regions or on the borders of an utter desert, will competition cease. The land may be extremely cold or dry, yet there will be competition between some few species, or between the individuals of the same species, for the warmest or dampest spots.

Hence, also, we can see that when a plant or animal is placed in a new country amongst new competitors, though the climate may be exactly the same as in its former home, yet the conditions of its life will generally be changed in an essential manner. If we wished to increase its average numbers in its new home, we should have to modify it in a different way to what we should have done in its native country; for we should have to give it some advantage over a different set of competitors or enemies.

It is good thus to try in our imagination to give any form some advantage over another. Probably in no single instance should we know what to do, so as to succeed. It will convince us of our ignorance on the mutual relations of all organic beings; a conviction as necessary, as it seems to be difficult to acquire. All that we can do, is to keep steadily in mind that each organic being is striving to increase at a geometrical ratio; that each at some period of its life, during some season of the year, during each generation or at intervals, has to struggle for life, and to suffer great destruction. When we reflect on this struggle, we may console ourselves with the full belief, that the war of nature is not incessant, that no fear is felt, that death is generally prompt, and that the vigorous, the healthy, and the happy survive and multiply.

Natural selection

How will the struggle for existence, discussed too briefly in the last chapter, act in regard to variation? Can the principle of selection, which we have seen[4] is so potent in the hands of man, apply in nature? I think we shall see that it can act most effectually. Let it be borne in mind in what an endless number of strange peculiarities our domestic productions, and, in a lesser degree, those under nature, vary; and how strong the hereditary tendency is. Under domestication, it may be truly said that the whole organisation becomes in some degree plastic. Let it be borne in mind how infinitely complex and close-fitting are the mutual relations of all organic beings to each other and to their physical conditions of life. Can it, then, be thought improbable, seeing that variations useful to man have undoubtedly occurred, that other variations useful in some way to each being in the great and complex battle of life, should sometimes occur in the course of thousands of generations? If such do occur, can we doubt (remembering that many more individuals are born than can possibly survive) that individuals having any advantage, however slight, over others, would have the best chance of surviving and of procreating their kind? On the other hand, we may feel sure that any variation in the least degree injurious would be rigidly destroyed. This preservation of favourable variations and the rejection of injurious variations, I call Natural Selection. Variations neither useful nor injurious would not be affected by natural selection, and would be left a fluctuating element, as perhaps we see in the species called polymorphic.[5]

We shall best understand the probable course of natural selection by taking the case of a country undergoing some physical change, for instance, of climate. The proportional numbers of its inhabitants would almost immediately undergo a change, and some species might become extinct. We may conclude, from what we have seen of the intimate and complex manner in which the inhabitants of each country are bound together, that any change in the numerical proportions of some of the inhabitants, independently of the change of climate itself, would most seriously affect many of the others. If the country were open on its borders, new forms would certainly immigrate, and this also would seriously disturb the relations of some of the former inhabitants. Let it be remembered how powerful the influence of a single introduced tree or mammal has been shown to be.[6] But in the case of an island, or of a country partly surrounded by barriers, into which new and better adapted forms could not freely enter, we should then have places in the economy of nature which would assuredly be better filled up, if some of the original inhabitants were in some manner modified; for, had the area been open to immigration, these same places would have been seized on by intruders. In such case, every slight modification, which in the course of ages chanced to arise, and which in any way favoured the individuals of any of the species, by better adapting them to their altered conditions, would tend to be preserved; and natural selection would thus have free scope for the work of improvement.

We have reason to believe . . . that a change in the conditions of life, by specially acting on the reproductive system, causes or increases variability;[7] and in the foregoing case the conditions of life are supposed to have undergone a change, and this would manifestly be favourable to natural selection, by giving a better chance of profitable variations occurring; and unless profitable variations do occur, natural selection can do nothing. Not that, as I believe, any extreme amount of variability is necessary; as man can certainly produce great results by adding up in any given direction mere individual differences, so could Nature, but far more easily, from having incomparably longer time at her disposal. Nor do I believe that any great physical change, as of climate, or any unusual degree of isolation to check immigration, is actually necessary to produce new and unoccupied places for natural selection to fill up by modifying and improving some of the varying inhabitants. For as all the inhabitants of each country are struggling together with nicely balanced forces, extremely slight modifications in the structure or habits of one inhabitant would often give it an advantage over others; and still further modifications of the same kind would often still further increase the advantage. No country can be named in which all the native inhabitants are now so perfectly adapted to each other and to the physical conditions under which they live, that none of them could anyhow be improved; for in all countries, the natives have been so far conquered by naturalised productions, that they have allowed foreigners to take firm possession of the land. And as foreigners have thus everywhere beaten some of the natives, we may safely conclude that the natives might have been modified with advantage, so as to have better resisted such intruders.

As man can produce and certainly has produced a great result by his methodical and unconscious means of selection, what may not nature effect? Man can act only on external and visible characters: nature cares nothing for appearances, except in so far as they may be useful to any being. She can act on every internal organ, on every shade of constitutional difference, on the whole machinery of life. Man selects only for his own good; Nature only for that of the being which she tends. Every selected character is fully exercised by her; and the being is placed under well-suited conditions of life. Man keeps the natives of many climates in the same country; he seldom exercises each selected character in some peculiar and fitting manner; he feeds a long and a short beaked pigeon on the same food; he does not exercise a long-backed or long-legged quadruped in any peculiar manner; he exposes sheep with long and short wool to the same climate. He does not allow the most vigorous males to struggle for the females. He does not rigidly destroy all inferior animals, but protects during each varying season, as far as lies in his power, all his productions. He often begins his selection by some half-monstrous form; or at least by some modification prominent enough to catch his eye, or to be plainly useful to him. Under nature, the slightest difference of structure or constitution may well turn the nicely-balanced scale in the struggle for life, and so be preserved. How fleeting are the wishes and efforts of man! how short his time! and consequently how poor will his products be, compared with those accumulated by nature during whole geological periods. Can we wonder, then, that nature's productions should be far 'truer' in character than man's productions; that they should be infinitely better adapted to the most complex conditions of life, and should plainly bear the stamp of far higher workmanship?

It may be said that natural selection is daily and hourly scrutinising, throughout the world, every variation, even the slightest; rejecting that which is bad, preserving and adding up all that is good; silently and insensibly working, whenever and wherever opportunity offers, at the improvement of each organic being in relation to its organic and inorganic conditions of life. We see nothing of these slow changes in progress, until the hand of time has marked the long lapses of ages, and then so imperfect is our view into long past geological ages, that we only see that the forms of life are now different from what they formerly were.

Although natural selection can act only through and for the good of each being, yet characters and structures, which we are apt to consider as of very trifling importance, may thus be acted on. When we see leaf-eating insects green, and bark-feeders mottled-grey; the alpine ptarmigan white in winter, the red-grouse the colour of heather, and the black-grouse that of peaty earth, we must believe that these tints are of service to these birds and insects in preserving them from danger. Grouse, if not destroyed at some period of their lives, would increase in countless numbers; they are known to suffer largely from birds of prey; and hawks are guided by eyesight to their prey,—so much so, that on parts of the Continent persons are warned not to keep white pigeons, as being the most liable to destruction. Hence I can see no reason to doubt that natural selection might be most effective in giving the proper colour to each kind of grouse, and in keeping that colour, when once acquired, true and constant. Nor ought we to think that the occasional destruction of an animal of any particular colour would produce little effect: we should remember how essential it is in a flock of white sheep to destroy every lamb with the faintest trace of black. In plants the down on the fruit and the colour of the flesh are considered by botanists as characters of the most trifling importance: yet we hear from an an excellent horticulturist, Downing, that in the United States smooth-skinned fruits suffer far more from a beetle, a *Curculio*, than those with down; that purple plums suffer far more from a certain disease than yellow plums; whereas another disease attacks yellow-fleshed peaches far more than those with other coloured flesh. If, with all the aids of art, these slight differences make a great difference in cultivating the several varieties, assuredly, in a state of nature, where the trees would have to struggle with other trees and with a host of enemies, such differences would effectually settle which variety, whether a smooth or downy, a yellow or purple fleshed fruit, should succeed.

In looking at many small points of difference between species, which, as far as our ignorance permits us to judge, seem to be quite unimportant, we must not forget that climate, food, &c., probably produce some slight and direct effect. It is, however, far more necessary to bear in mind that there are many unknown laws of correlation of growth, which, when one part of the organisation is modified through variation, and the modifications are accumulated by natural selection for the good of the being, will cause other modifications, often of the most unexpected nature.

As we see that those variations which under domestication appear at any particular period of life, tend to reappear in the offspring at the same period;—for instance, in the seeds of the many varieties of our culinary and agricultural plants; in the caterpillar and cocoon stages of the varieties of the silkworm; in the eggs of poultry, and in the colour of the down of their chickens; in the horns of our sheep and cattle when nearly adult;—so in a state of nature, natural selection will be enabled to act on and modify organic beings at any age, by the accumulation of profitable variations at that age, and by their inheritance at a corresponding age. If it profit a plant to have its seeds more and more widely disseminated by the wind, I can see no greater difficulty in this being effected through natural selection, than in the cotton-planter increasing and improving by selection the down in the pods on his cotton-trees. Natural selection may modify and adapt the larva of an insect to a score of contingencies, wholly different from those which concern the mature insect. These modifications will no doubt affect, through the laws of correlation, the structure of the adult; and probably in the case of those insects which live only for a few hours, and which never feed, a large part of their structure is merely the correlated result of successive changes in the structure of their larvae. So, conversely, modifications in the adult will probably often affect the structure of the larva; but in all cases natural selection will ensure that modifications consequent on other modifications at a different period of life, shall not be in the least degree injurious: for if they became so, they would cause the extinction of the species.

Natural selection will modify the structure of the young in relation to the parent, and of the parent in relation to the young. In social animals it will adapt the structure of each individual for the benefit of the community; if each in consequence profits by the selected change. What natural selection cannot do, is to modify the structure of one species, without giving it any advantage, for the good of another species; and though statements to this effect may be found in works of natural history, I cannot find one case which will bear investigation. A structure used only once in an animal's whole life, if of high importance to it, might be modified to any extent by natural selection; for instance, the great jaws possessed by certain insects, and used exclusively for opening the cocoon—or the hard tip to the beak of nestling birds, used for breaking the egg. It has been asserted, that of the best short-beaked tumbler-pigeons more perish in the egg than are able to get out of it; so that fanciers assist in the act of hatching. Now, if nature had to make the beak of a full-grown pigeon very short for the bird's own advantage, the process of modification would be very slow, and there would be simultaneously the most rigorous selection of the young birds within the egg, which had the most powerful and hardest beaks, for all with weak beaks would inevitably perish: or, more delicate and more easily broken shells might be selected, the thickness of the shell being known to vary like every other structure.

. . .

Illustrations of the action of Natural Selection. In order to make it clear how, as I believe, natural selection acts, I must beg permission to give one or two imaginary illustrations. Let us take the case of a wolf, which preys on various animals, securing some by craft, some by strength, and some by fleetness; and let us suppose that the fleetest prey, a deer for instance, had from any change in the country increased in numbers, or that other prey had decreased in numbers, during that season of the year when the wolf is hardest pressed for food. I can under such circumstances see no reason to doubt that the swiftest and slimmest wolves would have the best chance of surviving, and so be preserved or selected,—provided always that they retained strength to master their prey at this or at some other period of the year, when they might be compelled to prey on other animals. I can see no more reason to doubt this, than that man can improve the fleetness of his greyhounds by careful and methodical selection, or by that unconscious selection which results from each man trying to keep the best dogs without any thought of modifying the breed.

Even without any change in the proportional numbers of the animals on which our wolf preyed, a cub might be born with an innate tendency to pursue certain kinds of prey. Nor can this be thought very improbable; for we often observe great differences in the natural tendencies of our domestic animals; one cat, for instance, taking to catch rats, another mice; one cat, according to Mr. St. John, bringing home winged game, another hares or rabbits, and another hunting on marshy ground and almost nightly catching woodcocks or snipes. The tendency to catch rats rather than mice is known to be inherited. Now, if any slight innate change of habit or of structure benefited an individual wolf, it would have the best chance of surviving and of leaving offspring. Some of its young would probably inherit the same habits or structure, and by the repetition of this process, a new variety might be formed which would either supplant or coexist with the parent-form of wolf. Or, again, the wolves inhabiting a mountainous district, and those frequenting the low-lands, would naturally be forced to hunt different prey; and from the continued preservation of the individuals best fitted for the two sites, two varieties might slowly be formed. These varieties would cross and blend where they met; but to this subject of intercrossing we shall soon have to return. I may add, that, according to Mr. Pierce, there are two varieties of the wolf inhabiting the Catskill Mountains in the United States, one with a light greyhound-like form, which pursues deer, and the other more bulky, with shorter legs, which more frequently attacks the shepherd's flocks.

Let us now take a more complex case. Certain plants excrete a sweet juice, apparently for the sake of eliminating something injurious from their sap: this is effected by glands at the base of the stipules in some Leguminosae, and at the back of the leaf of the common laurel. This juice, though small in quantity, is greedily sought by insects. Let us now suppose a little sweet juice or nectar to be excreted by the inner bases of the petals of a flower. In this case insects in seeking the nectar would get dusted with pollen, and would certainly often transport the pollen from one flower to the stigma of another flower. The flowers of two distinct individuals of the same species would thus get crossed; and the act of crossing, we have good reason to believe,[8] ... would produce very vigorous seedlings, which consequently would have the best chance of flourishing and surviving. Some of these seedlings would probably inherit the nectar-excreting power. Those individual flowers which had the largest glands or nectaries, and which excreted most nectar, would be oftenest visited by insects, and would be oftenest crossed; and so in the long-run would gain the upper hand. Those flowers, also, which had their stamens and pistils placed, in relation to the size and habits of the particular insects which visited them, so as to favour in any degree the transportal of their pollen from flower to flower, would likewise be favoured or selected. We might have taken the case of insects visiting flowers for the sake of collecting pollen instead of nectar; and as pollen is formed for the sole object of fertilisation, its destruction appears a simple loss to the plant; yet if a little pollen were carried, at first occasionally and then habitually, by the pollen-devouring insects from flower to flower, and a cross thus effected, although nine-tenths of the pollen were destroyed, it might still be a great gain to the plant; and those individuals which produced more and more pollen, and had larger and larger anthers, would be selected.

When our plant, by this process of the continued preservation or natural selection of more and more attractive flowers, had been rendered highly attractive to insects, they would, unintentionally on their part, regularly carry pollen from flower to flower; and that they can most effectually do this, I could easily show by many striking instances. I will give only one—not as a very striking case, but as likewise illustrating one step in the separation of the sexes of plants, presently to be alluded to. Some holly-trees bear only male flowers, which have four stamens producing rather a small quantity of pollen, and a rudimentary pistil;[9] other holly-trees bear only female flowers; these have a full-sized pistil, and four stamens with shrivelled anthers, in which not a grain of pollen can be detected. Having found a female tree exactly sixty yards from a male tree, I put the stigmas[10] of twenty flowers, taken from different branches, under the microscope, and on all, without exception, there were pollen-grains, and on some a profusion of pollen. As the wind had set for several days from the female to the male tree, the pollen could not thus have been carried. The weather had been cold and boisterous, and therefore not favourable to bees, nevertheless every female flower which I examined had been effectually fertilised by the bees, accidentally dusted with pollen, having flown from tree to tree in search of nectar. But to return to our imaginary case: as soon as the plant had been rendered so highly attractive to insects that pollen was regularly carried from flower to flower, another process might commence. No naturalist doubts the advantage of what has been called the 'physiological divison of labour'; hence we may believe that it would be advantageous to a plant to produce stamens alone in one flower or on one whole plant, and pistils alone in another flower or on another plant. In plants under culture and placed under new conditions of life, sometimes the male organs and sometimes the female organs become more or less impotent; now if we suppose this to occur in ever so slight a degree under nature, then as pollen is already carried regularly from flower to flower, and as a more complete separation of the sexes of our plant would be advantageous on the principle of the division of labour, individuals with this tendency more and more increased, would be continually favoured or selected, until at last a complete separation of the sexes would be effected.

Let us now turn to the nectar-feeding insects in our imaginary case: we may suppose the plant of which we have been slowly increasing the nectar by continued selection, to be a common plant; and that certain insects depended in main part on its nectar for food. I could give many facts, showing how anxious bees are to save time; for instance, their habit of cutting holes and sucking the nectar at the bases of certain flowers, which they can, with a very little more trouble, enter by the mouth. Bearing such facts in mind, I can see no reason to doubt that an accidental deviation in the size and form of the body, or in the curvature and length of the proboscis, &c., far too slight to be appreciated by us, might profit a bee or other insect, so that an individual so characterised would be able to obtain its food more quickly, and so have a better chance of living and leaving descendants. Its descendants would probably inherit a tendency to a similar slight deviation of structure. The tubes of the corollas of the common red and incarnate clovers (*Trifolium pratense* and *incarnatum*) do not on a hasty glance appear to differ in length; yet the hive-bee can easily suck the nectar out of the incarnate clover, but not out of the common red clover, which is visited by humble-bees alone; so that whole fields of the red clover offer in vain an abundant supply of precious nectar to the hive-bee. Thus it might be a great advantage to the hive-bee to have a slightly longer or differently constructed proboscis. On the other hand, I have found by experiment that the fertility of clover greatly depends on bees visiting and moving parts of the corolla, so as to push the pollen on to the stigmatic surface. Hence, again, if humble-bees were to become rare in any country, it might be a great advantage to the red clover to have a shorter or more deeply divided tube to its corolla, so that the hive-bee could visit its flowers. Thus I can understand how a flower and a bee might slowly become, either simultaneously or one after the other, modified and adapted in the most perfect manner to each other, by the continued preservation of individuals presenting mutual and slightly favourable deviations of structure.

I am well aware that this doctrine of natural selection, exemplified in the above imaginary instances, is open to the same objections which were at first urged against Sir Charles Lyell's noble views on 'the modern changes of the earth, as illustrative of geology'; but we now very seldom hear the action, for instance, of the coast-waves, called a trifling and insignificant cause, when applied to the excavation of gigantic valleys or to the formation of the longest lines of inland cliffs.[11] Natural selection can act only by the preservation and accumulation of infinitesimally small inherited modifications, each profitable to the preserved being; and as modern geology has almost banished such views as the excavation of a great valley by a single diluvial wave, so will natural selection, if it be a true principle, banish the belief of the continued creation of new organic beings, or of any great and sudden modification in their structure.

. . .

Extinction. This subject . . . must be here alluded to from being intimately connected with natural selection. Natural selection acts solely through the preservation of variations in some way advantageous, which consequently endure. But as from the high geometrical powers of increase of all organic beings, each area is already fully stocked with inhabitants, it follows that as each selected and favoured form increases in number, so will the less favoured forms decrease and become rare. Rarity, as geology tells us, is the precursor to extinction. We can, also, see that any form represented by few individuals will, during fluctuations in the seasons or in the number of its enemies, run a good chance of utter extinction. But we may go further than this; for as new forms are continually and slowly being produced, unless we believe that the number of specific forms goes on perpetually and almost indefinitely increasing, numbers inevitably must become extinct. That the number of specific forms has not indefinitely increased, geology shows us plainly; and indeed we can see reason why they should not have thus increased, for the number of places in the polity of nature is not indefinitely great,—not that we have any means of knowing that any one region has as yet got its maximum of species. Probably no region is as yet fully stocked, for at the Cape of Good Hope, where more species of plants are crowded together than in any other quarter of the world, some foreign plants have become naturalised, without causing, as far as we know, the extinction of any natives.

Furthermore, the species which are most numerous in individuals will have the best chance of producing within any given period favourable variations. We have evidence of this, in the fact . . . that it is the common species which afford the greatest number of recorded varieties, or incipient species. Hence, rare species will be less quickly modified or improved within any given period, and they will consequently be beaten in the race for life by the modified descendants of the commoner species.

From these several considerations I think it inevitably follows, that as new species in the course of time are formed through natural selection, others will become rarer and rarer, and finally extinct. The forms which stand in closest competition with those undergoing modification and improvement, will naturally suffer most. And we have seen in the chapter on the Struggle for Existence that it is the most closely-allied forms,—varieties of the same species, and species of the same genus or of related genera,—which, from having nearly the same structure, constitution, and habits, generally come into the severest competition with each other. Consequently, each new variety or species, during the progress of its formation, will generally press hardest on its nearest kindred, and tend to exterminate them. We see the same process of extermination amongst our domesticated productions, through the selection of improved forms by man. Many curious instances could be given showing how quickly new breeds of cattle, sheep, and other animals, and varieties of flowers, take the place of older and inferior kinds. In Yorkshire, it is historically known that the ancient black cattle were displaced by the long-horns, and that these 'were swept away by the short-horns' (I quote the words of an agricultural writer) 'as if by some murderous pestilence.'

Divergence of Character.   The principle, which I have designated by this term, is of high importance on my theory, and explains, as I believe, several important facts. In the first place, varieties, even strongly-marked ones, though having somewhat of the character of species—as is shown by the hopeless doubts in many cases how to rank them—yet certainly differ from each other far less than do good and distinct species. Nevertheless, according to my view, varieties are species in the process of formation, or are, as I have called them incipient species. How, then, does the lesser difference between varieties become augmented into the greater difference between species? That this does habitually happen, we must infer from most of the innumerable species throughout nature presenting well-marked differences; whereas varieties, the supposed prototypes and parents of future well-marked species, present slight and ill-defined differences. Mere chance, as we may call it, might cause one variety to differ in some character from its parents, and the offspring of this variety again to differ from its parent in the very same character and in a greater degree; but this alone would never account for so habitual and large an amount of difference as that between varieties of the same species and species of the same genus.

As has always been my practice, let us seek light on this head from our domestic productions. We shall here find something analogous. A fancier is struck by a pigeon having a slightly shorter beak; another fancier is struck by a pigeon having a rather longer beak; and on the acknowledged principle that 'fanciers do not and will not admire a medium standard, but like extremes,' they both go on (as has actually occurred with tumbler-pigeons) choosing and breeding from birds with longer and longer beaks, or with shorter and shorter beaks. Again, we may suppose that at an early period one man preferred swifter horses; another stronger and more bulky horses. The early differences would be very slight; in the course of time, from the continued selection of swifter horses by some breeders, and of stronger ones by others, the differences would become greater, and would be noted as forming two sub-breeds; finally, after the lapse of centuries, the sub-breeds would become converted into two well-established and distinct breeds. As the differences slowly become greater, the inferior animals with intermediate characters, being neither very swift nor very strong, will have been neglected, and will have tended to disappear. Here, then, we see in man's productions the action of what may be called the principle of divergence, causing differences, at first barely appreciable, steadily to increase, and the breeds to diverge in character both from each other and from their common parent.

But how, it may be asked, can any analogous principle apply in nature? I believe it can and does apply most efficiently, from the simple circumstance that the more diversified the descendants from any one species become in structure, constitution, and habits, by so much will they be better enabled to seize on many and widely diversified places in the polity of nature, and so be enabled to increase in numbers.

We can clearly see this in the case of animals with simple habits. Take the case of a carnivorous quadruped, of which the number that can be supported in any country has long ago arrived at its full average. If its natural powers of increase be allowed to act, it can succeed in increasing (the country not undergoing any change in its conditions) only by its varying descendants seizing on places at present occupied by other animals: some of them, for instance, being enabled to feed on new kinds of prey, either dead or alive; some inhabiting new stations, climbing trees, frequenting water, and some perhaps becoming less carnivorous. The more diversified in habits and structure the descendants of our carnivorous animal became, the more places they would be enabled to occupy. What applies to one animal will apply throughout all time to all animals—that is, if they vary—for otherwise natural selection can do nothing. So it will be with plants. It has been experimentally proved, that if a plot of ground be sown with one species of grass, and a similar plot be sown with several distinct genera of grasses, a greater number of plants and a greater weight of dry herbage can thus be raised. The same has been found to hold good when first one variety and then several mixed varieties of wheat have been sown on equal spaces of ground. Hence, if any one species of grass were to go on varying, and those varieties were continually selected which differed from each other in at all the same manner as distinct species and genera of grasses differ from each other, a greater number of individual plants of this species of grass, including its modified descendants, would succeed in living on the same piece of ground. And we well know that each species and each variety of grass is annually sowing almost countless seeds; and thus, as it may be said, is striving its utmost to increase its numbers. Consequently, I cannot doubt that in the course of many thousands of generations, the most distinct varieties of any one species of grass would always have the best chance of succeeding and of increasing in numbers, and thus of supplanting the less distinct varieties; and varieties, when rendered very distinct from each other, take the rank of species.

The truth of the principle, that the greatest amount of life can be supported by great diversification of structure, is seen under many natural circumstances. In an extremely small area, especially if freely open to immigration, and where the contest between individual and individual must be severe, we always find great diversity in its inhabitants. For instance, I found that a piece of turf, three feet by four in size, which had been exposed for many years to exactly the same conditions, supported twenty species of plants, and these belonged to eighteen genera and to eight orders, which shows how much these plants differed from each other. So it is with the plants and insects on small and uniform islets; and so in small ponds of fresh water. Farmers find that they can raise most food by a rotation of plants belonging to the most different orders: nature follows what may be called a simultaneous rotation. Most of the animals and plants which live close round any small piece of ground, could live on it (supposing it not to be in any way peculiar in its nature), and may be said to be striving to the utmost to live there; but, it is seen, that where they come into the closest competition with each other, the advantages of diversification of structure, with the accompanying differences of habit and constitution, determine that the inhabitants, which thus jostle each other most closely, shall, as a general rule, belong to what we call different genera and orders.

The same principle is seen in the naturalisation of plants through man's agency in foreign lands. It might have been expected that the plants which have succeeded in becoming naturalised in any land would generally have been closely allied to the indigenes; for these are commonly looked at as specially created and adapted for their own country. It might, also, perhaps have been expected that naturalised plants would have belonged to a few groups more especially adapted to certain stations in their new homes. But the case is very different; and Alph. De Candolle[12] has well remarked in his great and admirable work, that floras gain by naturalisation, proportionally with the number of the native genera and species, far more in new genera than in new species. To give a single instance: in the last edition of Dr Asa Gray's 'Manual of the Flora of the Northern United States,' 260 naturalised plants are enumerated, and these belong to 162 genera. We thus see that these naturalised plants are of a highly diversified nature. They differ, moreover, to a large extent from the indigenes, for out of the 162 genera, no less than 100 genera are not there indigenous, and thus a large proportional addition is made to the genera of these States.

By considering the nature of the plants or animals which have struggled successfully with the indigenes of any country, and have there become naturalised, we can gain some crude idea in what manner some of the natives would have had to be modified, in order to have gained an advantage over the other natives; and we may, I think, at least safely infer that diversification of structure, amounting to new generic differences, would have been profitable to them.

The advantage of diversification in the inhabitants of the same region is, in fact, the same as that of the physiological division of labour in the organs of the same individual body—a subject so well elucidated by Milne Edwards.[13] No physiologist doubts that a stomach by being adapted to digest vegetable matter alone, or flesh alone, draws most nutriment from these substances. So in the general economy of any land, the more widely and perfectly the animals and plants are diversified for different habits of life, so will a greater number of individuals be capable of there supporting themselves. A set of animals, with their organisation but little diversified, could hardly compete with a set more perfectly diversified in structure. It may be doubted, for instance, whether the Australian marsupials, which are divided into groups differing but little from each other, and feebly representing, as Mr Waterhouse and others have remarked, our carnivorous, ruminant, and rodent mammals, could successfully compete with these well-pronounced orders. In the Australian mammals, we see the process of diversification in an early and incomplete stage of development.

After the foregoing discussion, which ought to have been much amplified, we may, I think, assume that the modified descendants of any one species will succeed by so much the better as they become more diversified in structure, and are thus enabled to encroach on places occupied by other beings. Now let us see how this principle of great benefit being derived from divergence of character, combined with the principles of natural selection and of extinction, will tend to act.

The accompanying diagram [Fig. 4.1] will aid us in understanding this rather perplexing subject. Let A to L represent the species of a genus large in its own country; these species are supposed to resemble each other in unequal degrees, as is so generally the case in nature, and as is represented in the diagram by the letters standing at unequal distances. I have said a large genus, because we have seen in the second chapter,[14] that on an average more of the species of large genera vary than of small genera; and the varying species of the large genera present a greater number of varieties. We have, also, seen that the species, which are the commonest and the most widely-diffused, vary more than rare species with restricted ranges. Let (A) be a

97

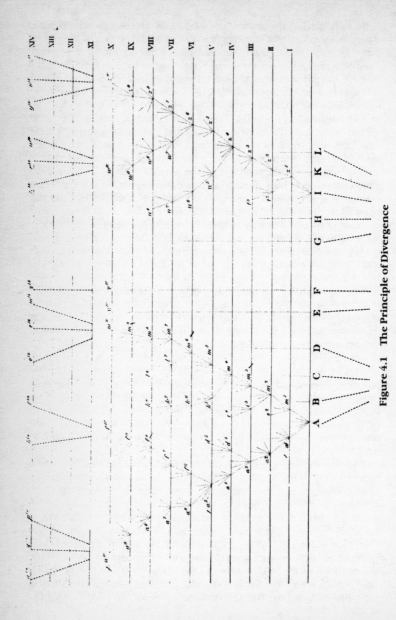

Figure 4.1 The Principle of Divergence

common, widely-diffused, and varying species, belonging to a genus
large in its own country. The little fan of diverging dotted lines of
unequal lengths proceeding from (A), may represent its varying
offspring. The variations are supposed to be extremely slight, but of
the most diversified nature; they are not supposed all to appear
simultaneously, but often after long intervals of time; nor are they all
supposed to endure for equal periods. Only those variations which
are in some way profitable will be preserved or naturally selected.
And here the importance of the principle of benefit being derived
from divergence of character comes in; for this will generally lead to
the most different or divergent variations (represented by the outer
dotted lines) being preserved and accumulated by natural selection.
When a dotted line reaches one of the horizontal lines, and is there
marked by a small numbered letter, a sufficient amount of variation is
supposed to have been accumulated to have formed a fairly
well-marked variety, such as would be thought worthy of record in a
systematic work.

The intervals between the horizontal lines in the diagram, may
represent each a thousand generations; but it would have been better
if each had represented ten thousand generations. After a thousand
generations, species (A) is supposed to have produced two fairly
well-marked varieties, namely a^1 and m^1. These two varieties will
generally continue to be exposed to the same conditions which made
their parents variable, and the tendency to variability is in itself
hereditary, consequently they will tend to vary, and generally to vary
in nearly the same manner as their parents varied. Moreover, these
two varieties, being only slightly modified forms, will tend to inherit
those advantages which made their common parent (A) more
numerous than most of the other inhabitants of the same country;
they will likewise partake of those more general advantages which
made the genus to which the parent-species belonged, a large genus
in its own country. And these circumstances we know to be
favourable to the production of new varieties.

If, then, these two varieties be variable, the most divergent of their
variations will generally be preserved during the next thousand
generations. And after this interval, variety a^1 is supposed in the
diagram to have produced variety a^2, which will, owing to the
principle of divergence, differ more from (A) than did variety a^1.
Variety m^1 is supposed to have produced two varieties, namely m^2
and s^2, differing from each other, and more considerably from their
common parent (A). We may continue the process by similar steps
for any length of time; some of the varieties, after each thousand
generations, producing only a single variety, but in a more and more
modified condition, some producing two or three varieties, and some

failing to produce any. Thus the varieties or modified descendants, proceeding from the common parent (A), will generally go on increasing in number and diverging in character. In the diagram the process is represented up to the ten-thousandth generation, and under a condensed and simplified form up to the fourteen-thousandth generation.

But I must here remark that I do not suppose that the process ever goes on so regularly as is represented in the diagram, though in itself made somewhat irregular. I am far from thinking that the most divergent varieties will invariably prevail and multiply: a medium form may often long endure, and may or may not produce more than one modified descendant; for natural selection will always act according to the nature of the places which are either unoccupied or not perfectly occupied by other beings; and this will depend on infinitely complex relations. But as a general rule, the more diversified in structure the descendants from any one species can be rendered, the more places they will be enabled to seize on, and the more their modified progeny will be increased. In our diagram the line of succession is broken at regular intervals by small numbered letters marking the successive forms which have become sufficiently distinct to be recorded as varieties. But these breaks are imaginary, and might have been inserted anywhere, after intervals long enough to have allowed the accumulation of a considerable amount of divergent variation.

As all the modified descendants from a common and widely-diffused species, belonging to a large genus, will tend to partake of the same advantages which made their parent successful in life, they will generally go on multiplying in number as well as diverging in character: this is represented in the diagram by the several divergent branches proceeding from (A). The modified offspring from the later and more highly improved branches in the lines of descent, will, it is probable, often take the place of, and so destroy, the earlier and less improved branches: this is represented in the diagram by some of the lower branches not reaching to the upper horizontal lines. In some cases I do not doubt that the process of modification will be confined to a single line of descent, and the number of the descendants will not be increased; although the amount of divergent modification may have been increased in the successive generations. This case would be represented in the diagram, if all the lines proceeding from (A) were removed, excepting that from a^1 to a^{10}. In the same way, for instance, the English race-horse and English pointer have apparently both gone on slowly diverging in character from their original stocks, without either having given off any fresh branches or races.

After ten thousand generations, species (A) is supposed to have produced three forms, a^{10}, f^{10}, and m^{10}, which, from having diverged in character during the successive generations, will have come to differ largely, but perhaps unequally, from each other and from their common parent. If we suppose the amount of change between each horizontal line in our diagram to be excessively small, these three forms may still be only well-marked varieties; or they may have arrived at the doubtful category of sub-species; but we have only to suppose the steps in the process of modification to be more numerous or greater in amount, to convert these three forms into well-defined species: thus the diagram illustrates the steps by which the small differences distinguishing varieties are increased into the larger differences distinguishing species. By continuing the same process for a greater number of generations (as shown in the diagram in a condensed and simplified manner), we get eight species, marked by the letters between a^{14} and m^{14}, all descended from (A). Thus, as I believe, species are multiplied and genera are formed.

In a large genus it is probable that more than one species would vary. In the diagram I have assumed that a second species (I) has produced, by analogous steps, after ten thousand generations, either two well-marked varieties (w^{10} and z^{10}) or two species, according to the amount of change supposed to be represented between the horizontal lines. After fourteen thousand generations, six new species, marked by the letters n^{14} to z^{14}, are supposed to have been produced. In each genus, the species, which are already extremely different in character, will generally tend to produce the greatest number of modified descendants; for these will have the best chance of filling new and widely different places in the polity of nature: hence in the diagram I have chosen the extreme species (A), and the nearly extreme species (I), as those which have largely varied, and have given rise to new varieties and species. The other nine species (marked by capital letters) of our original genus, may for a long period continue transmitting unaltered descendants; and this is shown in the diagram by the dotted lines not prolonged far upwards from want of space.

But during the process of modification, represented in the diagram, another of our principles, namely that of extinction, will have played an important part. As in each fully stocked country natural selection necessarily acts by the selected form having some advantage in the struggle for life over other forms, there will be a constant tendency in the improved descendants of any one species to supplant and exterminate in each stage of descent their predecessors and their original parent. For it should be remembered that the competition will generally be most severe between those forms which are most nearly related to each other in habits, constitution, and structure.

Hence all the intermediate forms between the earlier and later states, that is between the less and more improved state of a species, as well as the original parent-species itself, will generally tend to become extinct. So it probably will be with many whole collateral lines of descent, which will be conquered by later and improved lines of descent. If, however, the modified offspring of a species get into some distinct country, or become quickly adapted to some quite new station, in which child and parent do not come into competition, both may continue to exist.

If then our diagram be assumed to represent a considerable amount of modification, species (A) and all the earlier varieties will have become extinct, having been replaced by eight new species (a^{14} to m^{14}); and (I) will have been replaced by six (n^{14} to z^{14}) new species.

But we may go further than this. The original species of our genus were supposed to resemble each other in unequal degrees, as is so generally the case in nature; species (A) being more nearly related to B, C, and D, than to the other species; and species (I) more to G, H, K, L, than to the others. These two species (A) and (I), were also supposed to be very common and widely diffused species, so that they must originally have had some advantage over most of the other species of the genus. Their modified descendants, fourteen in number at the fourteen-thousandth generation, will probably have inherited some of the same advantages: they have also been modified and improved in a diversified manner at each stage of descent, so as to have become adapted to many related places in the natural economy of their country. It seems, therefore, to me extremely probable that they will have taken the places of, and thus exterminated, not only their parents (A) and (I), but likewise some of the original species which were most nearly related to their parents. Hence very few of the original species will have transmitted offspring to the fourteen-thousandth generation. We may suppose that only one (F), of the two species which were least closely related to the other nine original species, has transmitted descendants to this late stage of descent.

The new species in our diagram descended from the original eleven species, will now be fifteen in number. Owing to the divergent tendency of natural selection, the extreme amount of difference in character between species a^{14} and z^{14} will be much greater than that between the most different of the original eleven species. The new species, moreover, will be allied to each other in a widely different manner. Of the eight descendants from (A) the three marked a^{14}, q^{14}, p^{14}, will be nearly related from having recently branched off from a^{10}; b^{14} and f^{14}, from having diverged at an earlier period from a^5, will be in some degree distinct from the three first-named species; and lastly,

o^{14}, e^{14}, and m^{14}, will be nearly related one to the other, but from having diverged at the first commencement of the process of modification, will be widely different from the other five species, and may constitute a sub-genus or even a distinct genus.

The six descendants from (I) will form two sub-genera or even genera. But as the original species (I) differed largely from (A), standing nearly at the extreme points of the original genus, the six descendants from (I) will, owing to inheritance, differ considerably from the eight descendants from (A); the two groups, moreover, are supposed to have gone on diverging in different directions. The intermediate species, also (and this is a very important consideration), which connected the original species (A) and (I), have all become, excepting (F), extinct, and have left no descendants. Hence the six new species descended from (I), and the eight descended from (A), will have to be ranked as very distinct genera, or even as distinct sub-families.

Thus it is, as I believe, that two or more genera are produced by descent, with modification, from two or more species of the same genus. And the two or more parent-species are supposed to have descended from some one species of an earlier genus. In our diagram, this is indicated by the broken lines, beneath the capital letters, converging in sub-branches downwards towards a single point; this point representing a single species, the supposed single parent of our several new sub-genera and genera.

It is worth while to reflect for a moment on the character of the new species F^{14}, which is supposed not to have diverged much in character, but to have retained the form of (F), either unaltered or altered only in a slight degree. In this case, its affinities to the other fourteen new species will be of a curious and circuitous nature. Having descended from a form which stood between the two parent-species (A) and (I), now supposed to be extinct and unknown, it will be in some degree intermediate in character between the two groups descended from these species. But as these two groups have gone on diverging in character from the type of their parents, the new species (F^{14}) will not be directly intermediate between them, but rather between types of the two groups; and every naturalist will be able to bring some such case before his mind.

In the diagram, each horizontal line has hitherto been supposed to represent a thousand generations, but each may represent a million or hundred million generations, and likewise a section of the successive strata of the earth's crust including extinct remains. We shall, when we come to our chapter on Geology,[15] have to refer again to this subject, and I think we shall then see that the diagram throws light on the affinities of extinct beings, which, though generally belonging to the same orders, or families, or genera, with those now living, yet are often, in some degree, intermediate in character between existing groups; and we can understand this fact, for the extinct species lived at very ancient epochs when the branching lines of descent had diverged less.

I see no reason to limit the process of modification, as now explained, to the formation of genera alone. If, in our diagram, we suppose the amount of change represented by each successive group of diverging dotted lines to be very great, the forms marked a^{14} to p^{14}, those marked b^{14} and f^{14}, and those marked o^{14} to m^{14}, will form three very distinct genera. We shall also have two very distinct genera descended from (I); and as these latter two genera, both from continued divergence of character and from inheritance from a different parent, will differ widely from the three genera descended from (A), the two little groups of genera will form two distinct families, or even orders, according to the amount of divergent modification supposed to be represented in the diagram. And the two new families, or orders, will have descended from two species of the original genus; and these two species are supposed to have descended from one species of a still more ancient and unknown genus.

We have seen that in each country it is the species of the larger genera which oftenest present varieties or incipient species. This, indeed, might have been expected; for as natural selection acts through one form having some advantage over other forms in the struggle for existence, it will chiefly act on those which already have some advantage; and the largeness of any group shows that its species have inherited from a common ancestor some advantage in common. Hence, the struggle for the production of new and modified descendants, will mainly lie between the larger groups, which are all trying to increase in number. One large group will slowly conquer another large group, reduce its numbers, and thus lessen its chance of further variation and improvement. Within the same large group, the later and more highly perfected sub-groups, from branching out and seizing on many new places in the polity of Nature, will constantly tend to supplant and destroy the earlier and less improved sub-groups. Small and broken groups and sub-groups will finally tend to disappear. Looking to the future, we can predict that the groups of organic beings which are now large and triumphant, and which are least broken up, that is, which as yet have suffered least extinction, will for a long period continue to increase. But which groups will ultimately prevail, no man can predict; for we well know that many groups, formerly most extensively developed, have now become extinct. Looking still more remotely to the future, we may predict that, owing to the continued and steady increase of the larger groups, a multitude of smaller groups will become utterly extinct, and leave no modified descendants; and consequently that of the species living at any one period, extremely few will transmit descendants to a remote futurity. I shall have to return to this subject in the chapter on Classification,[16] but I may add that on this view of extremely few of the more ancient species having transmitted descendants, and on the view of all the descendants of the same species making a class, we can understand how it is that there exist but very few classes in each main division of the animal and vegetable kingdoms. Although extremely few of the most ancient species may now have living and modified descendants, yet at the most remote geological period, the earth may have been as well peopled with many species of many genera, families, orders, and classes, as at the present day.

Recapitulation

The recapitulatory chapter deals with the whole work, and we join it after Darwin has recapitulated the main difficulties for his theory, and has summarised the theory of natural selection itself. He now turns to the evidence for evolution – which he calls 'descent with modification' – by natural selection, as opposed to the separate creation of species.

I have already recapitulated, as fairly as I could, the opposed difficulties and objections: now let us turn to the special facts and arguments in favour of the theory.

On the view that species are only strongly marked and permanent varieties, and that each species first existed as a variety, we can see why it is that no line of demarcation can be drawn between species, commonly supposed to have been produced by special acts of creation, and varieties which are acknowledged to have been produced by secondary laws. On this same view we can understand how it is that in each region where many species of a genus have been produced, and where they now flourish, these same species should present many varieties; for where the manufactory of species has been active, we might expect, as a general rule, to find it still in action; and this is the case if varieties be incipient species. Moreover, the species of the large genera, which afford the greater number of varieties or incipient species, retain to a certain degree the character of varieties; for they differ from each other by a less amount of difference than do the species of smaller genera. The closely allied species also of the larger genera apparently have restricted ranges, and they are clustered in little groups round other species—in which respects they resemble varieties. These are strange relations on the view of each species having been independently created, but are intelligible if all species first existed as varieties.

As each species tends by its geometrical ratio of reproduction to increase inordinately in number; and as the modified descendants of each species will be enabled to increase by so much the more as they become more diversified in habits and structure, so as to be enabled to seize on many and widely different places in the economy of nature, there will be a constant tendency in natural selection to preserve the most divergent offspring of any one species. Hence during a long-continued course of modification, the slight differences, characteristic of varieties of the same species, tend to be augmented into the greater differences characteristic of species of the same genus. New and improved varieties will inevitably supplant and exterminate the older, less improved and intermediate varieties; and thus species are rendered to a large extent defined and distinct objects. Dominant species belonging to the larger groups tend to give birth to new and dominant forms; so that each large group tends to become still larger, and at the same time more divergent in character. But as all groups cannot thus succeed in increasing in size, for the world would not hold them, the more dominant groups beat the less dominant. This tendency in the large groups to go on increasing in size and diverging in character, together with the almost inevitable contingency of much extinction, explains the arrangement of all the forms of life, in groups subordinate to groups, all within a few great classes, which we now see everywhere around us, and which has prevailed throughout all time. This grand fact of the grouping of all organic beings seems to me utterly inexplicable on the theory of creation.

As natural selection acts solely by accumulating slight, successive, favourable variations, it can produce no great or sudden modification; it can act only by very short and slow steps. Hence the canon of 'Natura non facit saltum', which every fresh addition to our knowledge tends to make more strictly correct, is on this theory simply intelligible. We can plainly see why nature is prodigal in variety, though niggard in innovation. But why this should be a law of nature if each species has been independently created, no man can explain.

Many other facts are, as it seems to me, explicable on this theory. How strange it is that a bird, under the form of woodpecker, should have been created to prey on insects on the ground; that upland geese, which never or rarely swim, should have been created with webbed feet; that a thrush should have been created to dive and feed on sub-aquatic insects; and that a petrel should have been created with habits and structure fitting it for the life of an auk or grebe! and so on in endless other cases. But on the view of each species constantly trying to increase in number, with natural selection always ready to adapt the slowly varying descendants of each to any unoccupied or ill-occupied place in nature, these facts cease to be strange, or perhaps might even have been anticipated.

As natural selection acts by competition, it adapts the inhabitants of each country only in relation to the degree of perfection of their associates; so that we need feel no surprise at the inhabitants of any one country, although on the ordinary view supposed to have been specially created and adapted for that country, being beaten and supplanted by the naturalised productions from another land. Nor ought we to marvel if all the contrivances in nature be not, as far as we can judge, absolutely perfect; and if some of them be abhorrent to our ideas of fitness. We need not marvel at the sting of the bee causing the bee's own death; at drones being produced in such vast numbers for one single act, and being then slaughtered by their sterile sisters; at the astonishing waste of pollen by our fir-trees; at the instinctive hatred of the queen bee for her own fertile daughters; at ichneumonidae feeding within the live bodies of caterpillars; and at other such cases. The wonder indeed is, on the theory of natural selection, that more cases of the want of absolute perfection have not been observed.

The complex and little known laws governing variation are the same, as far as we can see, with the laws which have governed the producton of so-called specific forms. In both cases physical conditions seem to have produced but little direct effect; yet when varieties enter any zone, they occasionally assume some of the characters of the species proper to that zone. In both varieties and species, use and disuse seem to have produced some effect; for it is difficult to resist this conclusion when we look, for instance, at the logger-headed duck, which has wings incapable of flight, in nearly the same condition as in the domestic duck; or when we look at the burrowing tucutucu,[17] which is occasionally blind, and then at certain moles, which are habitually blind and have their eyes covered with skin; or when we look at the blind animals inhabiting the dark caves of America and Europe. In both varieties and species correction of growth seems to have played a most important part, so that when one part has been modified other parts are necessarily modified. In both varieties and species reversions to long-lost characters occur. How inexplicable on the theory of creation is the occasional appearance of stripes on the shoulder and legs of the several species of the horse-genus and in their hybrids! How simply is this fact explained if we believe that these species have descended from a

107

striped progenitor, in the same manner as the several domestic breeds of pigeon have descended from the blue and barred rock-pigeon!

On the ordinary view of each species having been independently created, why should the specific characters, or those by which the species of the same genus differ from each other, be more variable than the generic characters in which they all agree? Why, for instance, should the colour of a flower be more likely to vary in any one species of a genus, if the other species, supposed to have been created independently, have differently coloured flowers, than if all the species of the genus have the same coloured flowers? If species are only well-marked varieties, of which the characters have become in a high degree permanent, we can understand this fact; for they have already varied since they branched off from a common progenitor in certain characters, by which they have come to be specifically distinct from each other; and therefore these same characters would be more likely still to be variable than the generic characters which have been inherited without change for an enormous period. It is inexplicable on the theory of creation why a part developed in a very unusual manner in any one species of a genus, and therefore, as we may naturally infer, of great importance to the species, should be eminently liable to variation; but, on my view, this part has undergone, since the several species branched off from a common progenitor, an unusual amount of variability and modification, and therefore we might expect this part generally to be still variable. But a part may be developed in the most unusual manner, like the wing of a bat, and yet not be more variable than any other structure, if the part be common to many subordinate forms, that is, if it has been inherited for a very long period; for in this case it will have been rendered constant by long-continued natural selection.

Glancing at instincts, marvellous as some are, they offer no greater difficulty than does corporeal structure on the theory of the natural selection of successive, slight, but profitable modifications. We can thus understand why nature moves by graduated steps in endowing different animals of the same class with their several instincts. I have attempted to show how much light the principle of gradation throws on the admirable architectural powers of the hive-bee.[18] Habit no doubt sometimes comes into play in modifying instincts; but it certainly is not indispensable, as we see, in the case of neuter insects, which leave no progeny to inherit the effects of long-continued habit.[19] On the view of all the species of the same genus having descended from a common parent, and having inherited much in common, we can understand how it is that allied species, when placed under considerably different conditions of life, yet should follow nearly the same instincts; why the thrush of South America, for instance, lines her nest with mud like our British species. On the view of instincts having been slowly acquired through natural selection we need not marvel at some instincts being apparently not perfect and liable to mistakes, and at many instincts causing other animals to suffer.

If species be only well-marked and permanent varieties, we can at once see why their crossed offspring should follow the same complex laws in their degrees and kinds of resemblance to their parents,—in being absorbed into each other by successive crosses, and in other such points,—as do the crossed offspring of acknowledged varieties.[20] On the other hand, these would be strange facts if species have been independently created, and varieties have been produced by secondary laws.

If we admit that the geological record is imperfect in an extreme degree, then such facts as the record gives, support the theory of descent with modification. New species have come on the stage slowly and at successive intervals; and the amount of change, after equal intervals of time, is widely different in different groups. The extinction of species and of whole groups of species, which has played so conspicuous a part in the history of the organic world, almost inevitably follows on the principle of natural selection; for old forms will be supplanted by new and improved forms. Neither single species nor groups of species reappear when the chain of ordinary generation has once been broken. The gradual diffusion of dominant forms, with the slow modification of their descendants, causes the forms of life, after long intervals of time, to appear as if they had changed simultaneously throughout the world. The fact of the fossil remains of each formation being in some degree intermediate in character between the fossils in the formations above and below, is simply explained by their intermediate position in the chain of descent. The grand fact that all extinct organic beings belong to the same system with recent beings, falling either into the same or into intermediate groups, follows from the living and the extinct being the offspring of common parents. As the groups which have descended from an ancient progenitor have generally diverged in character, the progenitor with its early descendants will often be intermediate in character in comparison with its later descendants; and thus we can see why the more ancient a fossil is, the oftener it stands in some degree intermediate between existing and allied groups. Recent forms are generally looked at as being, in some vague sense, higher than ancient and extinct forms; and they are in so far higher as the later and more improved forms have conquered the older and less

The origin of species

improved organic beings in the struggle for life. Lastly, the law of the long endurance of allied forms on the same continent,—of marsupials in Australia, of edentata[21] in America, and other such cases,—is intelligible, for within a confined country, the recent and the extinct will naturally be allied by descent.

Looking to geographical distribution, if we admit that there has been during the long course of ages much migration from one part of the world to another, owing to former climatal and geographical changes and to the many occasional and unknown means of dispersal, then we can understand, on the theory of descent with modification, most of the great leading facts in Distribution. We can see why there should be so striking a parallelism in the distribution of organic beings throughout space, and in their geological succession throughout time; for in both cases the beings have been connected by the bond of ordinary generation, and the means of modification have been the same. We see the full meaning of the wonderful fact, which must have struck every traveller, namely, that on the same continent, under the most diverse conditions, under heat and cold, on mountain and lowland, on deserts and marshes, most of the inhabitants within each great class are plainly related; for they will generally be descendants of the same progenitors and early colonists. On this same principle of former migration, combined in most cases with modification, we can understand, by the aid of the Glacial period, the identity of some few plants, and the close alliance of many others, on the most distant mountains, under the most different climates; and likewise the close alliance of some of the inhabitants of the sea in the northern and southern temperate zones, though separated by the whole intertropical ocean. Although two areas may present the same physical conditions of life, we need feel no surprise at their inhabitants being widely different, if they have been for a long period completely separated from each other; for as the relation of organism to organism is the most important of all relations, and as the two areas will have received colonists from some third source or from each other, at various periods and in different proportions, the course of modification in the two areas will inevitably be different.

On this view of migration, with subsequent modification, we can see why oceanic islands should be inhabited by few species, but of these, that many should be peculiar. We can clearly see why those animals which cannot cross wide spaces of ocean, as frogs and terrestrial mammals, should not inhabit oceanic islands; and why, on the other hand, new and peculiar species of bats, which can traverse the ocean, should so often be found on islands far distant from any continent. Such facts as the presence of peculiar species of bats, and the absence of all other mammals, on oceanic islands, are utterly

110

inexplicable on the theory of independent acts of creation.

The existence of closely allied or representative species in any two areas, implies, on the theory of descent with modification, that the same parents formerly inhabited both areas; and we almost invariably find that wherever many closely allied species inhabit two areas, some identical species common to both still exist. Wherever many closely allied yet distinct species occur, many doubtful forms and varieties of the same species likewise occur. It is a rule of high generality that the inhabitants of each area are related to the inhabitants of the nearest source whence immigrants might have been derived. We see this in nearly all the plants and animals of the Galapagos archipelago, of Juan Fernandez, and of the other American islands being related in the most striking manner to the plants and animals of the neighbouring American mainland; and those of the Cape de Verde archipelago and other African islands to the African mainland. It must be admitted that these facts receive no explanation on the theory of creation.

The fact, as we have seen, that all past and present organic beings constitute one grand natural system[22] with group subordinate to group, and with extinct groups often falling in between recent groups, is intelligible on the theory of natural selection with its contingencies of extinction and divergence of character. On these same principles we see how it is, that the mutual affinities of the species and genera within each class are so complex and circuitous. We see why certain characters are far more serviceable than others for classification;—why adaptive characters, though of paramount importance to the being, are of hardly any importance in classification; why characters derived from rudimentary parts, though of no service to the being, are often of high classificatory value; and why embryological characters are the most valuable of all. The real affinities of all organic beings are due to inheritance or community of descent. The natural system is a genealogical arrangement, in which we have to discover the lines of descent by the most permanent characters, however slight their vital importance may be.

The framework of bones being the same in the hand of a man, wing of a bat, fin of the porpoise, and leg of the horse,—the same number of vertebrae forming the neck of the giraffe and of the elephant,—and innumerable other such facts, at once explain themselves on the theory of descent with slow and slight successive modifications. The similarity of pattern[23] in the wing and leg of a bat, though used for such different purposes,—in the jaws and legs of a crab,—in the petals, stamens, and pistils of a flower, is likewise intelligible on the view of the gradual modification of parts or organs, which were alike in the early progenitor of each class. On the principle of successive variations not always supervening at an early age, and being inherited at a corresponding not early period of life, we can clearly see why the embryos of mammals, birds, reptiles, and fishes should be so closely alike, and should be so unlike the adult forms. We may cease marvelling at the embryo of an air-breathing mammal or bird having branchial slits and arteries running in loops, like those in a fish which has to breathe the air dissolved in water, by the aid of well-developed branchiae.

Disuse, aided sometimes by natural selection, will often tend to reduce an organ, when it has become useless by changed habits or under changed conditions of life; and we can clearly understand on this view the meaning of rudimentary organs. But disuse and selection will generally act on each creature, when it has come to maturity and has to play its full part in the struggle for existence, and will thus have little power of acting on an organ during early life; hence the organ will not be much reduced or rendered rudimentary at this early age. The calf, for instance, has inherited teeth, which never cut through the gums of the upper jaw, from an early progenitor having well-developed teeth; and we may believe, that the teeth in the mature animal were reduced, during successive generations, by disuse or by the tongue and palate having been fitted by natural selection to browse without their aid; whereas in the calf, the teeth have been left untouched by selection or disuse, and on the principle of inheritance at corresponding ages have been inherited from a remote period to the present day. On the view of each organic being and each separate organ having been specially created, how utterly inexplicable it is that parts, like the teeth in the embryonic calf or like the shrivelled wings under the soldered wing-covers of some beetles, should thus so frequently bear the plain stamp of inutility! Nature may be said to have taken pains to reveal, by rudimentary organs and by homologous structures, her scheme of modification, which it seems that we wilfully will not understand.

Notes

1 The first two chapters had documented the fact of variability in domestic and natural plants and animals.

2 Augustin-Pyramus de Candolle, 1778–1841. Swiss botanist. Father of Alphonse de Candolle (see note 12 below). The 'future work' referred to in the previous sentence was never written.

3 Darwin was horrified when he found the sums were wrong. In the 6th edition, the numbers were 740 to 750 years, and 19 million elephants. The calculation is by no means trivial.

4 In Chapter 1 of the *Origin*, not included here. We must therefore take it for granted that artificial selection has produced striking results; as we shall see,

Darwin constantly explains natural selection by analogy with artificial selection by humans.

5   Polymorphism means the co-existence within a species of more than one discrete, recognisable form. The melanic and peppered types of the moth *Biston betularia* are a well-known example. Whether polymorphism is really due to 'variations neither useful nor injurious' – which would now be called neutral variation – is a matter of controversy.

6   i.e. on pp. 79–80 above.

7   The source of new inherited variants was then unknown. In Chapter 1, Darwin had quoted evidence that new variants arise when organisms are reared in altered conditions.

8   See Chapter 8 below.

9   The female reproductive part of the flower; the anther is the male part.

10   The stigma is the part of the female reproductive organ where the pollen is received.

11   On the importance of Lyell, see pp. 7–8.

12   Alphonse de Candolle, 1806–1893. Swiss botanist. The 'great and admirable work' is his *Géographique botanique raisonée* (1855).

13   Henri Milne-Edwards, 1800–1885. French zoologist.

14   Chapter 2 is not included in this anthology.

15   The geological chapter is not included in this anthology. The general point, though, is clear here and is spelled out again on p. 109.

16   The chapter on classification is not included in this anthology. The principle of divergence accounts for the pattern of similarity between species that are reflected in the way they are classified together.

17   South American burrowing rodents.

18   In the chapter on 'instinct', not included here, Darwin had considered how various kind of hives, made by various different species of bees, could have arisen in an evolutionary series.

19   The 'workers' of a colony of social insects (bees, wasps, ants, termites) do not reproduce and yet they come in a variety of forms within each species. These forms could not have arisen by the inheritance of acquired characteristics, because they do not reproduce, and thus rule out 'Lamarckism' as a general theory of evolution.

20   A variety is a distinct form of a species. If species had been separately created, but varieties arose naturally, why should crosses among varieties show the same laws of inheritance as crosses among species? Species generally do not interbreed, but can sometimes be made to; Darwin considered the results of 'hybridism' in a chapter of the *Origin*.

21   Anteaters, armadillos, sloths.

22   i.e. that living things can successfully be classified into a hierarchy of (for example, in the lion) genus (*Panthera*), within family (Felidae), within order (Carnivora), within class (Mammalia), within phylum (Chordata), within kingdom (Animalia).

23   Called homology.

—————— *Chapter five* ——————

The variation of animals and plants under domestication (1868)

At the beginning of 1860, two months after the publication of the *Origin*, Darwin (as he tells us in his *Autobiography*) 'began arranging my notes for my work on the Variation of Animals and Plants under domestication; but it was not published until the beginning of 1868; the delay having been partly caused by frequent illnesses, one of which lasted seven months, and partly by having been tempted to publish on other subjects which at the time interested me more . . . It is a big book and cost me four years and two months hard labour.' He described the two large volumes as follows: 'It gives all my observations and an immense number of facts collected from various sources about our domestic productions. In the second volume the causes and laws of variation, inheritance, &c., are discussed, as far as our present state of knowledge permits. Towards the end of the work I give my well-abused hypothesis of Pangenesis.' Pangenesis was Darwin's theory of inheritance, and his chapter on it is the subject of this extract. The chapter summarises much of the factual material of the earlier chapters and is the most famous chapter in the book. Above all, it is famous for being wrong.

However, for the theory of evolution – which was Darwin's

main interest – that did not matter. Darwin was well aware of this. Three of the first five chapters of the *Origin of species* were about inheritance and variation, but Darwin had not mentioned his theory of pangenesis there. And not because he had not invented it. He had been investigating inheritance ever since he had returned from the *Beagle* voyage and had probably formulated the provisional hypothesis by 1840–1. He had left it out of the *Origin* because it was not relevant. Natural selection did require that characters be inherited, and he showed that in fact they were; but to understand, beyond that, exactly how characters were passed on from one generation to the next was a secondary problem. Darwin had no great confidence in his theory of pangenesis and wished to avoid unnecessary controversy: he therefore held fire until he had time to write the larger, separate book.

Heredity, according to Darwin's pangenetic theory, is effected by small particles called gemmules. Throughout the life of an organism, all organs of its body (Darwin thought) were producing these gemmules, and the nature of the gemmules encapsulated the nature of the organ. A strong muscle produced gemmules for strong muscles; a weak one produced gemmules for weak muscles. The gemmules were released from their place of origin, circulated in the body, and aggregated in the gametes (sperms and eggs, pollen and ovules). They would then be passed on to the offspring and control the offspring's development. Now, as it happens, these hypothetical gemmules do not exist. The material (called DNA) in the gametes that actually does control inheritance is not influenced by the rest of the body; no hereditary factors pass to the gametes from the rest of the body. Characters acquired during the life of an organism are not passed on. It was a German biologist, August Weismann, who particularly persuaded biologists that the reproductive ('germ') and mortal ('soma') parts of a body are independent. Weismann first published his 'germ plasm' theory in 1883, a year after Darwin's death.

So Darwin's theory turned out to be wrong. It is, however, instructive and highly interesting. Darwin's argument resembles in style many of his other grand theories. Let us consider how he sets the argument out. The chapter has two parts. In the first part, he summarises (from earlier chapters) seven main classes of facts which a satisfactory theory of inheritance must explain: the variety of kinds of reproduction and their similarity to one another; graft hybrids; the influence of the male on the female reproductive system; development; the functional independence of the parts of the body; variability, including the effects of use and disuse (i.e. 'Lamarckian inheritance'); and inheritance, especially the inheritance of reversions. The meaning of all these terms will become clear in the extract (except for 'graft hybrids', which I have left out – 'Pangenesis does not throw much light on hybridism', Darwin wrote). At the beginning of the second part Darwin expounds pangenesis as 'a provisional hypothesis or speculation' capable of accounting for the facts and then goes on to show how it can explain each of the seven crucial classes of observations.

Pangenesis has many of the characteristics of a typical Darwinian theory. In it, Darwin posits an unobserved and hypothetical mechanism (gemmules); he collects all the evidence that is available, and shows how the hypothesis is consistent with it and can explain it better than can other hypotheses; he extrapolates the idea across the whole range of the subject; and he counters the imaginative problem of dealing with very large numbers (in this case, the very large numbers of gemmules produced by one body). In all these abstract respects, pangenesis resembles natural selection. Darwin's method let him down in the one case but not the other mainly because it relied on accumulations of anecdotal evidence. In the case of inheritance, the evidence was not reliable. Within the world of 19th-century biology, both the inheritance of acquired characters and the influence of males on the female reproductive system were, so to speak, 'facts the whole world knows'. They were almost universally accepted, and had never been properly tested. When they were, they were revealed to be untrue, and the theories that they supported collapsed. But that was to come after Darwin's death.

Darwin's theory does, if read in isolation, appear very different from the modern science of genetics. Modern genetics is

Weismannist and Mendelian, chromosomal and molecular. A modern genetics text-book will not mention Darwin's theory. But the theory was influential – and beneficially so – in the history of genetics. It did not influence Mendel; but, in the last three decades of the 19th century, it was discussed much more than any other theory of heredity. Few people accepted it, but everyone had to consider it. The most immediate need was to bring the theory up to date. In 1868 gemmules already had an old-fashioned air: they did not fit in with the cell-theory. The cell-theory demanded that cells arise from cells; and yet, if Darwin was right, gametic cells arose from aggregations of gemmules, which had in turn been produced by many different cells. It was as an attempt to reconcile pangenesis and the cell-theory that the Dutch geneticist De Vries, one of the re-discoverers of Mendelism, was led to his theory of 'intracellular pangenesis'. The other re-discoverers of Mendelism, and Weismann too, had also all grappled with Darwin's theory. The quotation from Whewell that Darwin put on the first page of the chapter therefore turned out to be remarkably prescient. Pangenesis was an error, but a scientifically valuable error. Thus the pangenesis chapter is still of interest both as an illustration of Darwin's scientific method and for its historical importance.

The extract that follows reproduces most of the pangenesis chapter. I have left in many of Darwin's back-references to his earlier chapters, even though they are not included in the anthology; the back-references give some idea of the contents of the earlier chapters of *Variation under domestication*. Unlike the other extracts in this anthology, I have cut out several sections – perhaps a third of the chapter in all – from within the chapter: I have removed some of the description of the pangenetic mechanism itself, and some of the factual explorations; but (with the exception of hybrids, mentioned above) I have retained a fairly complete skeleton of Darwin's argument.

Provisional hypothesis of pangenesis

In the previous chapters large classes of facts, such as those bearing on bud-variation, the various forms of inheritance, the causes and laws of variation, have been discussed; and it is obvious that these subjects, as well as the several modes of reproduction, stand in some sort of relation to each other. I have been led, or rather forced, to form a view which to a certain extent connects these facts by a tangible method. Every one would wish to explain to himself, even in an imperfect manner, how it is possible for a character possessed by some remote ancestor suddenly to reappear in the offspring; how the effects of increased or decreased use of a limb can be transmitted to the child; how the male sexual element can act not solely on the ovule, but occasionally on the mother-form; how a limb can be reproduced on the exact line of amputation, with neither too much nor too little added; how it comes that organic beings identical in every respect are habitually produced by such widely different processes, as budding and true seminal generation. I am aware that my view is merely a provisional hypothesis or speculation; but until a better one be advanced, it may be serviceable by bringing together a multitude of facts which are at present left disconnected by any efficient cause. As Whewell, the historian of the inductive sciences, remarks:—"Hypotheses may often be of service to science, when they involve a certain portion of incompleteness, and even of error." Under this point of view I venture to advance the hypothesis of Pangenesis, which implies that the whole organisation, in the sense of every separate atom or unit, reproduces itself. Hence ovules and pollen-grains,—the fertilised seed or egg, as well as buds,—include and consist of a multitude of germs thrown off from each separate atom of the organism.

In the First Part I will enumerate as briefly as I can the groups of facts which seem to demand connection; but certain subjects, not hitherto discussed, must be treated at disproportionate length. In the Second Part the hypothesis will be given; and we shall see, after considering how far the necessary assumptions are in themselves improbable, whether it serves to bring under a single point of view the various facts.

PART I.

Reproduction may be divided into two main classes, namely, sexual and asexual. The latter is effected in many ways—by gemmation, that is by the formation of buds of various kinds, and by fissiparous generation, that is by spontaneous or artificial division. It is notorious that some of the lower animals, when cut into many pieces, reproduce so many perfect individuals: Lyonnet cut a Nais or freshwater worm into nearly forty pieces, and these all reproduced perfect animals. It is probable that segmentation could be carried much further in some of the protozoa, and with some of the lowest plants each cell will reproduce the parent-form. Johannes Müller[1] thought that there was an important distinction between gemmation and fission; for in the latter case the divided portion, however small, is more perfectly organised; but most physiologists are now convinced that the two processes are essentially alike. Prof. Huxley remarks, "fission is little more than a peculiar mode of budding," and Prof. H. J. Clark, who has especially attended to this subject, shows in detail that there is sometimes "a compromise between self-division and budding." When a limb is amputated, or when the whole body is bisected, the cut extremities are said to bud forth; and as the papilla, which is first formed, consists of undeveloped cellular tissue like that forming an ordinary bud, the expression is apparently correct. We see the connection of the two processes in another way; for Trembley[2] observed that with the hydra the reproduction of the head after amputation was checked as soon as the animal began to bud.

Between the production, by fissiparous generation, of two or more complete individuals, and the repair of even a very slight injury, we have, as remarked in a former chapter, so perfect and insensible a gradation, that it is impossible to doubt that they are connected processes. Between the power which repairs a trifling injury in any part, and the power which previously "was occupied in its maintenance by the continued mutation of its particles", there cannot be any great difference; and we may follow Mr. Paget[3] in believing them to be the self-same power. As at each stage of growth an amputated part is replaced by one in the same state of development, we must likewise follow Mr. Paget in admitting "that the powers of development from the embryo are identical with those exercised for the restoration from injuries: in other words, that the powers are the same by which perfection is first achieved, and by which, when lost, it is recovered." Finally, we may conclude that the several forms of gemmation, and of fissiparous generation, the repair of injuries, the maintenance of each part in its proper state, and the growth or progressive development of the whole structure of the embryo, are all essentially the results of one and the same great power.

Sexual Generation. —The union of the two sexual elements seems to make a broad distinction between sexual and asexual reproduction. But the well-ascertained cases of Parthenogenesis prove that the distinction is not really so great as it at first appears; for ovules occasionally, and even in some cases frequently, become developed into perfect beings, without the concourse of the male element. J. Müller and others admit that ovules and buds have the same essential nature; and in the case of Daphnia Sir J. Lubbock[4] first showed that ova and pseudova are identical in structure. Certain bodies, which during their early development cannot be distinguished by any external character from true ovules, nevertheless must be classed as buds, for though formed within the ovarium they are incapable of fertilisation. This is the case with the germ-balls of the Cecidomyide larvæ,[5] as described by Leuckart. Ovules and the male element, before they become united, have, like buds, an independent existence. Both have the power of transmitting every single character possessed by the parent-form. We see this clearly when hybrids are paired *inter se*, for the characters of either grandparent often reappear, either perfectly or by segments, in the progeny. It is an error to suppose that the male transmits certain characters and the female other characters; though no doubt, from unknown causes, one sex sometimes has a stronger power of transmission than the other.

It has been maintained by some authors that a bud differs essentially from a fertilised germ, by always reproducing the perfect character of the parent-stock; whilst fertilised germs become developed into beings which differ, in a greater or less degree, from each other and from their parents. But there is no such broad distinction as this. In the eleventh chapter, numerous cases were given showing that buds occasionally grow into plants having new and strongly marked characters; and varieties thus produced can be propagated for a length of time by buds, and occasionally by seed. Nevertheless, it must be admitted that beings produced sexually are much more liable to vary than those produced asexually; and of this fact a partial explanation will hereafter be attempted. The variability in both cases is determined by the same general causes, and is governed by the same laws. Hence new varieties arising from buds cannot be distinguished from those arising from seed. Although bud-varieties usually retain their character during successive bud-generations, yet they occasionally revert, even after a long series of bud-generations, to their former character. This tendency to reversion in buds is one of the most remarkable of the several points of agreement between the offspring from bud and seminal reproduction.

. . .

From the several foregoing considerations we may conclude that the difference between sexual and asexual generation is not nearly so great as it at first appears; and we have already seen that there is the closest agreement between gemmation, fissiparous generation, the repair of injuries, and ordinary growth or development. The capacity of fertilisation by the male element seems to be the chief distinction between an ovule and a bud; and this capacity is not invariably brought into action, as in the cases of parthenogenetic reproduction. We are here naturally led to inquire what the final cause can be of the necessity in ordinary generation for the concourse of the two sexual elements.

Seeds and ova are often highly serviceable as the means of disseminating plants and animals, and of preserving them during one or more seasons in a dormant state; but unimpregnated seeds or ova, and detached buds, would be equally serviceable for both purposes. We can, however, indicate two important advantages gained by the concourse of the two sexes, or rather of two individuals belonging to opposite sexes; for, as I have shown in a former chapter, the structure of every organism appears to be especially adapted for the concurrence, at least occasionally, of two individuals. In nearly the same manner as it is admitted by naturalists that hybridism, from inducing sterility, is of service in keeping the forms of life distinct and fitted for their proper places; so, when species are rendered highly variable by changed conditions of life, the free intercrossing of the varying individuals will tend to keep each form fitted for its proper place in nature; and crossing can be effected only by sexual generation, but whether the end thus gained is of sufficient importance to account for the first origin of sexual intercourse is very doubtful. Secondly, I have shown, from the consideration of a large body of facts, that, as a slight change in the conditions of life is beneficial to each creature, so, in an analogous manner, is the change effected in the germ by sexual union with a distinct individual; and I have been led, from observing the many widely-extended provisions throughout nature for this purpose, and from the greater vigour of crossed organisms of all kinds, as proved by direct experiments, as well as from the evil effects of close interbreeding when long continued, to believe that the advantage thus gained is very great. Besides these two important ends, there may, of course, be others, as yet unknown to us, gained by the concourse of the two sexes.

. . .

Direct Action of the Male Element on the Female.—In the chapter just referred to, I have given abundant proofs that foreign pollen occasionally affects the mother-plant in a direct manner. Thus, when Gallesio fertilised an orange-flower with pollen from the lemon, the fruit bore stripes of perfectly characterised lemon-peel: with peas, several observers have seen the colour of the seed-coats and even of the pod directly affected by the pollen of a distinct variety; so it has been with the fruit of the apple, which consists of the modified calyx and upper part of the flower-stalk. These parts in ordinary cases are

121

wholly formed by the mother-plant. We here see the male element affecting and hybridising not that part which it is properly adapted to affect, namely the ovule, but the partially-developed tissues of a distinct individual. We are thus brought half-way towards a graft-hybrid, in which the cellular tissue of one form, instead of its pollen, is believed to hybridise the tissues of a distinct form. I formerly assigned reasons for rejecting the belief that the mother-plant is affected through the intervention of the hybridised embryo; but even if this view were admitted, the case would become one of graft-hybridism, for the fertilised embryo and the mother-plant must be looked at as distinct individuals.

With animals which do not breed until nearly mature, and of which all the parts are then fully developed, it is hardly possible that the male element should directly affect the female. But we have the analogous and perfectly well-ascertained case of the male element of a distinct form, as with the quagga and Lord Morton's mare, affecting the ovarium of the female, so that the ovules and offspring subsequently produced by her when impregnated by other males are plainly affected and hybridised by the first male.

Development.—The fertilised germ reaches maturity by a vast number of changes: these are either slight and slowly effected, as when the child grows into the man, or are great and sudden, as with the metamorphoses of most insects. Between these extremes we have, even within the same class, every gradation: thus, as Sir. J. Lubbock has shown, there is an Ephemerous insect[6] which moults above twenty times, undergoing each time a slight but decided change of structure; and these changes, as he further remarks, probably reveal to us the normal stages of development which are concealed and hurried through, or suppressed, in most other insects. In ordinary metamorphoses, the parts and organs appear to become changed into the corresponding parts in the next stage of development; but there is another form of development, which has been called by Professor Owen metagenesis. In this case "the new parts are not moulded upon the inner surface of the old ones. The plastic force has changed its course of operation. The outer case, and all that gave form and character to the precedent individual, perish and are cast off; they are not changed into the corresponding parts of the new individual. These are due to a new and distinct developmental process," &c. Metamorphosis, however, graduates so insensibly into metagenesis, that the two processes cannot be distinctly separated. For instance, in the last change which Cirripedes undergo, the alimentary canal and some other organs are moulded on pre-existing parts; but the eyes of the old and the young animal are developed in

122

entirely different parts of the body; the tips of the mature limbs are formed within the larval limbs, and may be said to be metamorphosed from them; but their basal portions and the whole thorax are developed in a plane actually at right angles to the limbs and thorax of the larva; and this may be called metagenesis. The metagenetic process is carried to an extreme degree in the development of some Echinoderms, for the animal in the second stage of development is formed almost like a bud within the animal of the first stage, the latter being then cast off like an old vestment, yet sometimes still maintaining for a short period an independent vitality.

If, instead of a single individual, several were to be thus developed metagenetically within a pre-existing form, the process would be called one of alternate generation. The young thus developed may either closely resemble the encasing parent-form, as with the larvæ of Cecidomyia, or may differ to an astonishing degree, as with many parasitic worms and with jelly-fishes; but this does not make any essential difference in the process, any more than the greatness or abruptness of the change in the metamorphoses of insects.

The whole question of development is of great importance for our present subject. When an organ, the eye for instance, is metageneti-cally formed in a part of the body where during the previous stage of development no eye existed, we must look at it as a new and independent growth. The absolute independence of new and old structures, which correspond in structure and function, is still more obvious when several individuals are formed within a previous encasing form, as in the cases of alternate generation. The same important principle probably comes largely into play even in the case of continuous growth, as we shall see when we consider the inheritance of modifications at corresponding ages.

. . .

The Functional Independence of the Elements or Units of the Body.—Physiologists agree that the whole organism consists of a multitude of elemental parts, which are to a great extent independent of each other. Each organ, says Claude Bernard,[7] has its proper life, its autonomy; it can develop and reproduce itself independently of the adjoining tissues. The great German authority, Virchow,[8] asserts still more emphatically that each system, as the nervous or osseous system, or the blood, consists of an "enormous mass of minute centres of action. . . . Every element has its own special action, and even though it derive its stimulus to activity from other parts, yet alone effects the actual performance of its duties. . . . Every single epithelial and muscular fibre-cell leads a sort of parasitical existence in relation to the rest of the body.... Every single bone-corpuscle really possesses conditions of nutrition peculiar to itself." Each element, as Mr. Paget remarks, lives its appointed time, and then dies, and, after being cast off or absorbed, is replaced. I presume that no physiologist doubts that, for instance, each bone-corpuscle of the finger differs from the corresponding corpuscle in the corresponding joint of the toe; and there can hardly be a doubt that even those on the corresponding sides of the body differ, though almost identical in nature. This near approach to identity is curiously shown in many diseases in which the same exact points on the right and left sides of the body are similarly affected; thus Mr. Paget gives a drawing of a diseased pelvis, in which the bone has grown into a most complicated pattern, but "there is not one spot or line on one side which is not represented, as exactly as it would be in a mirror, on the other."

Many facts support this view of the independent life of each minute element of the body. Virchow insists that a single bone-corpuscle or a single cell in the skin may become diseased. The spur of a cock, after being inserted into the eye of an ox, lived for eight years, and acquired a weight of 306 grammes, or nearly fourteen ounces. The tail of a pig has been grafted into the middle of its back, and reacquired sensibility. Dr. Ollier inserted a piece of periosteum from the bone of a young dog under the skin of a rabbit, and true bone was developed. A multitude of similar facts could be given. The frequent presence of hairs and of perfectly developed teeth, even teeth of the second dentition, in ovarian tumours, are facts leading to the same conclusion.

Whether each of the innumerable autonomous elements of the body is a cell or the modified product of a cell, is a more doubtful question, even if so wide a definition be given to the term, as to include cell-like bodies without walls and without nuclei. Professor Lionel Beale[9] uses the term "germinal matter" for the contents of cells, taken in this wide acceptation, and he draws a broad distinction between germinal matter and "formed material" or the various products of cells. But the doctrine of *omnis cellula e cellulâ* is admitted for plants, and is a widely prevalent belief with respect to animals. Thus Virchow, the great supporter of the cellular theory, whilst allowing that difficulties exist, maintains that every atom of tissue is derived from cells, and these from pre-existing cells, and these primarily from the egg, which he regards as a great cell. That cells, still retaining the same nature, increase by self-division or proliferation, is admitted by almost every one. But when an organism undergoes a great change of structure during development, the cells, which at each stage are supposed to be directly derived from previously existing cells, must likewise be greatly changed in nature;

124

this change is apparently attributed by the supporters of the cellular doctrine to some inherent power which the cells possess, and not to any external agency.

Another school maintains that cells and tissues of all kinds may be formed, independently of pre-existing cells, from plastic lymph or blastema; and this it is thought is well exhibited in the repair of wounds. As I have not especially attended to histology, it would be presumptuous in me to express an opinion on the two opposed doctrines. But every one appears to admit that the body consists of a multitude of "organic units", each of which possesses its own proper attributes, and is to a certain extent independent of all others. Hence it will be convenient to use indifferently the terms cells or organic units or simply units.

Variability and Inheritance.—We have seen in the twenty-second chapter that variability is not a principle co-ordinate with life or reproduction, but results from special causes, generally from changed conditions acting during successive generations. Part of the fluctuating variability thus induced is apparently due to the sexual system being easily affected by changed conditions, so that it is often rendered impotent; and when not so seriously affected, it often fails in its proper function of transmitting truly the characters of the parents to the offspring. But variability is not necessarily connected with the sexual system, as we see from the cases of bud-variation; and although we may not be able to trace the nature of the connexion, it is probable that many deviations of structure which appear in sexual offspring result from changed conditions acting directly on the organisation, independently of the reproductive organs. In some instances we may feel sure of this, when all, or nearly all the individuals which have been similarly exposed are similarly and definitely affected—as in the dwarfed and otherwise changed maize brought from hot countries when cultivated in Germany; in the change of the fleece in sheep within the tropics; to a certain extent in the increased size and early maturity of our highly-improved domesticated animals; in inherited gout from intemperance; and in many other such cases. Now, as such changed conditions do not especially affect the reproductive organs, it seems mysterious on any ordinary view why their product, the new organic being, should be similarly affected.

How, again, can we explain to ourselves the inherited effects of the use or disuse of particular organs? The domesticated duck flies less and walks more than the wild duck, and its limb-bones have become in a corresponding manner diminished and increased in comparison with those of the wild duck. A horse is trained to certain paces, and the colt inherits similar consensual movements. The domesticated rabbit becomes tame from close confinement; the dog intelligent from associating with man; the retriever is taught to fetch and carry: and these mental endowments and bodily powers are all inherited. Nothing in the whole circuit of physiology is more wonderful. How can the use or disuse of a particular limb or of the brain affect a small aggregate of reproductive cells, seated in a distant part of the body, in such a manner that the being developed from these cells inherits the characters of either one or both parents? Even an imperfect answer to this question would be satisfactory.

Sexual reproduction does not essentially differ, as we have seen, from budding or self-division, and these processes graduate through the repair of injuries into ordinary development and growth; it might therefore be expected that every character would be as regularly transmitted by all the methods of reproduction as by continued growth. In the chapters devoted to inheritance it was shown that a multitude of newly-acquired characters, whether injurious or beneficial, whether of the lowest or highest vital importance, are often faithfully transmitted—frequently even when one parent alone possesses some new peculiarity. It deserves especial attention that characters appearing at any age tend to reappear at a corresponding age. We may on the whole conclude that in all cases inheritance is the rule, and non-inheritance the anomaly. In some instances a character is not inherited, from the conditions of life being directly opposed to its development; in many instances, from the conditions incessantly inducing fresh variability, as with grafted fruit-trees and highly cultivated flowers. In the remaining cases the failure may be attributed to reversion, by which the child resembles its grand-parents or more remote progenitors, instead of its parents.

This principle of Reversion is the most wonderful of all the attributes of Inheritance. It proves to us that the transmission of a character and its development, which ordinarily go together and thus escape discrimination, are distinct powers; and these powers in some cases are even antagonistic, for each acts alternately in successive generations. Reversion is not a rare event, depending on some unusual or favourable combination of circumstances, but occurs so regularly with crossed animals and plants, and so frequently with uncrossed breeds, that it is evidently an essential part of the principle of inheritance. We know that changed conditions have the power of evoking long-lost characters, as in the case of some feral animals. The act of crossing in itself possesses this power in a high degree. What can be more wonderful than that characters, which have disappeared during scores, or hundreds, or even thousands of generations, should suddenly reappear perfectly developed, as in the case of pigeons and fowls when purely bred, and especially when crossed; or as with the zebrine stripes or dun-coloured horses, and other such cases? Many monstrosities come under this same head, as when rudimentary organs are redeveloped, or when an organ which we must believe was possessed by an early progenitor, but of which not even a rudiment is left, suddenly reappears, as with the fifth stamen in some Scrophulariaceæ. We have already seen that reversion acts in bud-reproduction; and we know that it occasionally acts during the growth of the same individual animal, especially, but not exclusively, when of crossed parentage,—as in the rare cases described of individual fowls, pigeons, cattle, and rabbits, which have reverted as they advanced in years to the colours of one of their parents or ancestors.

We are led to believe, as formerly explained, that every character which occasionally reappears is present in a latent form in each generation, in nearly the same manner as in male and female animals secondary characters of the opposite sex lie latent, ready to be evolved when the reproductive organs are injured. This comparison of the secondary sexual characters which are latent in both sexes, with other latent characters, is the more appropriate from the case recorded of the Hen, which assumed some of the masculine characters, not of her own race, but of an early progenitor; she thus exhibited at the same time the redevelopment of latent characters of both kinds and connected both classes. In every living creature we may feel assured that a host of lost characters lie ready to be evolved under proper conditions. How can we make intelligible, and connect with other facts, this wonderful and common capacity of reversion,— this power of calling back to life long-lost characters?

PART II.

I have now enumerated the chief facts which every one would desire to connect by some intelligible bond. This can be done, as it seems to me, if we make the following assumptions; if the first and chief one be not rejected, the others, from being supported by various physiological considerations, will not appear very improbable. It is almost universally admitted that cells, or the units of the body, propagate themselves by self-division or proliferation, retaining the same nature, and ultimately becoming converted into the various tissues and substances of the body. But besides this means of increase I assume that cells, before their conversion into completely passive or "formed material", throw off minute granules or atoms, which circulate freely throughout the system, and when supplied with proper nutriment multiply by self-division, subsequently becoming developed into cells like those from which they were derived. These granules for the sake of distinctness may be called cell-gemmules, or, as the cellular theory is not fully established, simply gemmules. They are supposed to be transmitted from the parents to the offspring, and are generally developed in the generation which immediately succeeds, but are often transmitted in a dormant state during many generations and are then developed. Their development is supposed to depend on their union with other partially developed cells or gemmules which precede them in the regular course of growth. Why I use the term union, will be seen when we discuss the direct action of pollen on the tissues of the mother-plant. Gemmules are supposed to be thrown off by every cell or unit, not only during the adult state, but during all the stages of development. Lastly, I assume that the gemmules in their dormant state have a mutual affinity for each other, leading to their aggregation either into buds or into the sexual elements. Hence, speaking strictly, it is not the reproductive elements, nor the buds, which generate new organisms, but the cells themselves throughout the body. These assumptions constitute the provisional hypothesis which I have called Pangenesis. Views in some respects similar have been propounded, as I find, by other authors.

Before proceeding to show, firstly, how far these assumptions are in themselves probable, and secondly, how far they connect and explain the various groups of facts with which we are concerned, it may be useful to give an illustration of the hypothesis. If one of the simplest Protozoa be formed, as appears under the microscope, of a small mass of homogeneous gelatinous matter, a minute atom thrown off from any part and nourished under favourable circumstances would naturally reproduce the whole; but if the upper and lower surfaces were to differ in texture from the central portion, then all three parts would have to throw off atoms or gemmules, which when aggregated by mutual affinity would form either buds or the sexual elements. Precisely the same view may be extended to one of the higher animals; although in this case many thousand gemmules must be thrown off from the various parts of the body. Now, when the leg, for instance, of a salamander is cut off, a slight crust forms over the wound, and beneath this crust the uninjured cells or units of bone, muscle, nerves, &c., are supposed to unite with the diffused gemmules of those cells which in the perfect leg come next in order; and these as they become slightly developed unite with others, and so on until a papilla of soft cellular tissue, the "budding leg", is formed, and in time a perfect leg. Thus, that portion of the leg which had been cut off, neither more nor less, would be reproduced. If the tail or leg of a young animal had been cut off, a young tail or leg would have been reproduced, as actually occurs with the amputated tail of the tadpole; for gemmules of all the units which compose the tail are diffused throughout the body at all ages. But during the adult state the gemmules of the larval tail would remain dormant, for they would not meet with pre-existing cells in a proper state of development with which to unite. If from changed conditions or any other cause any part of the body should become permanently modified, the gemmules, which are merely minute portions of the contents of the cells forming the part, would naturally reproduce the same modification. But gemmules previously derived from the same part before it had undergone any change, would still be diffused throughout the organisation, and would be transmitted from generation to generation, so that under favourable circumstances they might be redeveloped, and then the new modification would be for a time or for ever lost. The aggregation of gemmules derived from every part of the body, through their mutual affinity, would form buds, and their aggregation in some special manner, apparently in small quantity, together probably with the presence of gemmules of certain primordial cells, would constitute the sexual elements. By means of these illustrations the hypothesis of pangenesis has, I hope, been rendered intelligible.

Physiologists maintain, as we have seen, that each cell, though to a large extent dependent on others, is likewise, to a certain extent, independent or autonomous. I go one small step further, and assume that each cell casts off a free gemmule, which is capable of reproducing a similar cell. There is some analogy between this view and what we see in compound animals and in the flower-buds on the same tree; for these are distinct individuals capable of true or seminal reproduction, yet have parts in common and are dependent on each other; thus the tree has its bark and trunk, and certain corals, as the Virgularia, have not only parts, but movements in common.

The existence of free gemmules is a gratuitous assumption, yet can hardly be considered as very improbable, seeing that cells have the power of multiplication through the self-division of their contents. Gemmules differ from true ovules or buds inasmuch as they are supposed to be capable of multiplication in their undeveloped state. No one probably will object to this capacity as improbable. The blastema within the egg has been known to divide and give birth to two embryos; and Thuret has seen the zoospore of an alga divide itself, and both halves germinate. An atom of small-pox matter, so minute as to be borne by the wind, must multiply itself many thousandfold in a person thus inoculated. It has recently been ascertained that a minute portion of the mucous discharge from an animal affected with rinderpest, if placed in the blood of a healthy ox, increases so fast that in a short space of time "the whole mass of blood, weighing many pounds, is infected, and every small particle of that blood contains enough poison to give, within less than forty-eight hours, the disease to another animal."

The retention of free and undeveloped gemmules in the same body from early youth to old age may appear improbable, but we should remember how long seeds lie dormant in the earth and buds in the bark of a tree. Their transmission from generation to generation may appear still more improbable; but here again we should remember that many rudimentary and useless organs are transmitted and have been transmitted during an indefinite number of generations. We shall presently see how well the long continued transmission of undeveloped gemmules explains many facts.

As each unit, or group of similar units throughout the body, casts off its gemmules, and as all are contained within the smallest egg or seed, and within each spermatozoon or pollen-grain, their number and minuteness must be something inconceivable. I shall hereafter recur to this objection, which at first appears so formidable; but it may here be remarked that a cod-fish has been found to produce 6,867,840 eggs, a single Ascaris about 64,000,000 eggs, and a single Orchidaceous plant probably as many million seeds. In these several cases, the spermatozoa and pollen-grains must exist in considerably larger numbers. Now, when we have to deal with numbers such as these, which the human intellect cannot grasp, there is no good reason for rejecting our present hypothesis on account of the assumed existence of cell-gemmules a few thousand times more numerous.

. . .

The assumed elective affinity of each gemmule for that particular cell which precedes it in the order of development is supported by many analogies. In all ordinary cases of sexual reproduction the male and female elements have a mutual affinity for each other: thus, it is believed that about ten thousand species of Compositæ exist, and there can be no doubt that if the pollen of all these species could be, simultaneously or successively, placed on the stigma of any one species, this one would elect with unerring certainty its own pollen. This elective capacity is all the more wonderful, as it must have been acquired since the many species of this great group of plants branched off from a common progenitor. On any view of the nature of sexual reproduction, the protoplasm contained within the ovules and within the sperm-cells (or the "spermatic force" of the latter, if so vague a term be preferred) must act on each other by some law of special affinity, either during or subsequently to impregnation, so that corresponding parts alone affect each other; thus, a calf produced

130

from a short-horned cow by a long-horned bull has its horns and not its horny hoofs affected by the union of the two forms, and the offspring from two birds with differently coloured tails have their tails and not their whole plumage affected.

The various tissues of the body plainly show, as many physiologists have insisted, an affinity for special organic substances, whether natural or foreign to the body. We see this in the cells of the kidneys attracting urea from the blood; in the worrara poison affecting the nerves; upas and digitalis the muscles; the *Lytta vesicatoria* the kidneys;[10] and in the poisonous matter of many diseases, as small-pox, scarlet-fever, hooping-cough, glanders, cancer, and hydrophobia, affecting certain definite parts of the body or certain tissues or glands.

. . .

It has also been assumed that the development of each gemmule depends on its union with another cell or unit which has just commenced its development, and which, from preceding it in order of growth, is of a somewhat different nature. Nor is it a very improbable assumption that the development of a gemmule is determined by its union with a cell slightly different in nature, for abundant evidence was given in the seventeenth chapter, showing that a slight degree of differentiation in the male and female sexual elements favours in a marked manner their union and subsequent development. But what determines the development of the gem-mules of the first-formed or primordial cell in the unimpregnated ovule, is beyond conjecture.

. . .

Having now endeavoured to show that the several foregoing assumptions are to a certain extent supported by analogous facts, and having discussed some of the most doubtful points, we will consider how far the hypothesis brings under a single point of view the various cases ennumerated in the First Part. All the forms of reproduction graduate into each other and agree in their product; for it is impossible to distinguish between organisms produced from buds, from self-division, or from fertilised germs; such organisms are liable to variations of the same nature and to reversion of character; and as we now see that all the forms of reproduction depend on the aggregation of gemmules derived from the whole body, we can understand this general agreement. It is satisfactory to find that sexual and asexual generation, by both of which widely different processes the same living creature is habitually produced, are fundamentally the same. Parthenogenesis is no longer wonderful; in fact, the wonder is that it should not oftener occur. We see that the reproductive organs do not actually create the sexual elements; they merely determine or permit the aggregation of the gemmules in a special manner. . . . What determines the aggregation of the gemmules within the sexual organs we do not in the least know; nor do we know why buds are formed in certain definite places, leading to the symmetrical growth of trees and corals, nor why adventitious buds may be formed almost anywhere, even on a petal, and frequently upon healed wounds. As soon as the gemmules have aggregated themselves, development apparently commences, but in the case of buds is often afterwards suspended, and in the case of the sexual elements soon ceases, unless the elements of the opposite sexes combine; even after this has occurred, the fertilised germ, as with seeds buried in the ground, may remain during a lengthened period in a dormant state.

The antagonism which has long been observed, though exceptions occur, between active growth and the power of sexual reproduction—between the repair of injuries and gemmation—and with plants, between rapid increase by buds, rhizomes, &c., and the production of seed, is partly explained by the gemmules not existing in sufficient numbers for both processes. But this explanation hardly applies to those plants which naturally produce a multitude of seeds, but which, through a comparatively small increase in the number of the buds on their rhizomes or offsets, yield few or no seed. As, however, we shall presently see that buds probably include tissue which has already been to a certain extent developed or differentiated, some additional organised matter will thus have been expended.

From one of the forms of Reproduction, namely, spontaneous self-division, we are led by insensible steps to the repair of the slightest injury; and the existence of gemmules, derived from every cell or unit throughout the body and everywhere diffused, explains all such cases,—even the wonderful fact that, when the limbs of the salamander were cut off many times successively by Spallanzani and Bonnet,[11] they were exactly and completely reproduced. I have heard this process compared with the re-crystallisation which occurs when the angles of a broken crystal are repaired; and the two processes have this much in common, that in the one case the polarity of the molecules is the efficient cause, and in the other the affinity of the gemmules for particular nascent cells.

. . .

Abundant evidence has been advanced proving that pollen taken from one species or variety and applied to the stigma of another sometimes directly affects the tissues of the mother-plant. It is probable that this occurs with many plants during fertilisation, but can only be detected when distinct forms are crossed. On any ordinary theory of reproduction this is a most anomalous circumstance, for the pollen-grains are manifestly adapted to act on the ovule, but in these cases they act on the colour, texture, and form of the coats of the seeds, on the ovarium itself, which is a modified leaf, and even on the calyx and upper part of the flower-peduncle. In accordance with the hypothesis of pangenesis pollen includes gemmules, derived from every part of the organisation, which diffuse themselves and multiply by self-division: hence it is not surprising that gemmules within the pollen, which are derived from the parts near the reproductive organs, should sometimes be able to affect the same parts, whilst still undergoing development, in the mother plant.

As, during all the stages of development, the tissues of plants consist of cells, and as new cells are not known to be formed between, or independently of, pre-existing cells, we must conclude that the gemmules derived from the foreign pollen do not become developed merely in contact with pre-existing cells, but actually penetrate the nascent cells of the mother-plant. This process may be compared with the ordinary act of fertilisation, during which the contents of the pollen-tubes penetrate the closed embryonic sack within the ovule, and determine the development of the embryo. According to this view, the cells of the mother-plant may almost literally be said to be fertilised by the gemmules derived from the foreign pollen. With all organisms, as we shall presently see, the cells or organic units of the embryo during the successive stages of development may in like manner be said to be fertilised by the gemmules of the cells, which come next in the order of formation.

Animals, when capable of sexual reproduction, are fully developed, and it is scarcely possible that the male element should affect the tissues of the mother in the same direct manner as with plants; nevertheless it is certain that her ovaria are sometimes affected by a previous impregnation, so that the ovules subsequently fertilised by a distinct male are plainly influenced in character; and this, as in the case of foreign pollen, is intelligible through the diffusion, retention, and action of the gemmules included within the spermatozoa of the previous male.

Each organism reaches maturity through a longer or shorter course of development. The changes may be small and insensibly slow, as when a child grows into a man, or many, abrupt, and slight, as in the metamorphoses of certain ephemeral insects, or again few and strongly marked, as with most other insects. Each part may be moulded within a previously existing and corresponding part, and in this case it will appear, falsely as I believe, to be formed from the old part; or it may be developed within a wholly distinct part of the body, as in the extreme cases of metagenesis. An eye, for instance, may be developed at a spot where no eye previously existed. We have also seen that allied organic beings in the course of their metamorphoses sometimes attain nearly the same structure after passing through widely different forms; or conversely, after passing through nearly the same early forms, arrive at a widely different termination. In these cases it is very difficult to believe that the early cells or units possess the inherent power, independently of any external agent, of producing new structures wholly different in form, position, and function. But these cases become plain on the hypothesis of pangenesis. The organic units, during each stage of development, throw off gemmules, which, multiplying, are transmitted to the offspring. In the offspring, as soon as any particular cell or unit in the proper order of development becomes partially developed, it unites with (or to speak metaphorically is fertilised by) the gemmule of the next succeeding cell, and so onwards. Now, supposing that at any stage of development, certain cells or aggregates of cells had been slightly modified by the action of some disturbing cause, the cast-off gemmules or atoms of the cell-contents could hardly fail to be similarly affected, and consequently would reproduce the same modification. This process might be repeated until the structure of the part at this particular stage of development became greatly changed, but this would not necessarily affect other parts whether previously or subsequently developed. In this manner we can understand the remarkable independence of structure in the successive metamorphoses, and especially in the successive metageneses of many animals.

The term growth ought strictly to be confined to mere increase of size, and development to change of structure. Now, a child is said to grow into a man, and a foal into a horse, but, as in these cases there is much change of structure, the process properly belongs to the order of development. We have indirect evidence of this in many variations and diseases supervening during so-called growth at a particular period, and being inherited at a corresponding period. In the case, however, of diseases which supervene during old age, subsequently to the ordinary period of procreation, and which nevertheless are sometimes inherited, as occurs with brain and heart complaints, we must suppose that the organs were in fact affected at an earlier age and threw off at this period affected gemmules; but that the affection became visible or injurious only after the prolonged growth of the part in the strict sense of the word. In all the changes of structure which regularly supervene during old age, we see the effects of deteriorated growth, and not of true development.

. . .

134

The principle of the independent formation of each part, in so far as its development depends on the union of the proper gemmules with certain nascent cells, together with the superabundance of the gemmules derived from both parents and self-multiplied, throws light on a widely different group of facts, which on any ordinary view of development appears very strange. I allude to organs which are abnormally multiplied or transposed. Thus gold-fish often have supernumerary fins placed on various parts of their bodies. We have seen that, when the tail of a lizard is broken off, a double tail is sometimes reproduced, and when the foot of the salamander is divided longitudinally, additional digits are occasionally formed. When frogs, toads, &c., are born with their limbs doubled, as sometimes occurs, the doubling, as Gervais remarks, cannot be due to the complete fusion of two embryos, with the exception of the limbs, for the larvæ are limbless. The same argument is applicable to certain insects produced with multiple legs or antennæ, for these are metamorphosed from apodal or antennæless larvæ. Alphonse Milne-Edwards has described the curious case of a crustacean in which one eye-peduncle supported, instead of a complete eye, only an imperfect cornea, out of the centre of which a portion of an antenna was developed.

...

I do not know how physiologists look at such facts as the foregoing. According to the doctrine of pangenesis, the free and superabundant gemmules of the transposed organs are developed in the wrong place, from uniting with wrong cells or aggregates of cells during their nascent state; and this would follow from a slight modification in the elective affinity of such cells, or possibly of certain gemmules.

...

Variability often depends, as I have attempted to show, on the reproductive organs being injuriously affected by changed conditions; and in this case the gemmules derived from the various parts of the body are probably aggregated in an irregular manner, some superfluous and others deficient. Whether a superabundance of gemmules, together with fusion during development, would lead to the increased size of any part cannot be told; but we can see that their partial deficiency, without necessarily leading to the entire abortion of the part, might cause considerable modifications; for in the same manner as a plant, if its own pollen be excluded, is easily hybridised, so, in the case of a cell, if the properly succeeding gemmules were absent, it would probably combine easily with other and allied gemmules. We see this in the case of imperfect nails growing on the stumps of amputated fingers, for the gemmules of the nails have manifestly been developed at the nearest point.

In variations caused by the direction of changed conditions, whether of a definite or indefinite nature, as with the fleeces of sheep in hot countries, with maize grown in cold countries, with inherited gout, &c., the tissues of the body, according to the doctrine of pangenesis, are directly affected by the new conditions, and consequently throw off modified gemmules, which are transmitted with their newly acquired peculiarities to the offspring. On any ordinary view it is unintelligible how changed conditions, whether acting on the embryo, the young or adult animal, can cause inherited modifications. It is equally or even more unintelligible on any ordinary view, how the effects of the long-continued use or disuse of any part, or of changed habits of body or mind, can be inherited. A more perplexing problem can hardly be proposed; but on our view we have only to suppose that certain cells become at last not only functionally but structurally modified; and that these throw off similarly modified gemmules. This may occur at any period of development, and the modification will be inherited at a corresponding period; for the modified gemmules will unite in all ordinary cases with the proper preceding cells, and they will consequently be developed at the same period at which the modification first arose. With respect to mental habits or instincts, we are so profoundly ignorant on the relation between the brain and the power of thought that we do not know whether an inveterate habit or trick induces any change in the nervous system; but when any habit or other mental attribute, or insanity, is inherited, we must believe that some actual modification is transmitted; and this implies, according to our hypothesis, that gemmules derived from modified nerve-cells are transmitted to the offspring.

It is generally, perhaps always, necessary that an organism should be exposed during several generations to changed conditions or habits, in order that any modification in the structure of the offspring should ensue. This may be partly due to the changes not being at first marked enough to catch the attention, but this explanation is insufficient; and I can account for the fact, only by the assumption, which we shall see under the head of reversion is strongly supported, that gemmules derived from each cell before it had undergone the least modification are transmitted in large numbers to successive generations, but that the gemmules derived from the same cells after modification, naturally go on increasing under the same favouring conditions, until at last they become sufficiently numerous to overpower and supplant the old gemmules.

Another difficulty may be here noticed; we have seen that there is an important difference in the frequency, though not in the nature, of the variations in plants propagated by sexual and asexual generation. As far as variability depends on the imperfect action of the reproductive organs under changed conditions, we can at once see why seedlings should be far more variable than plants propagated by buds. We know that extremely slight causes,—for instance, whether a tree has been grafted or grows on its own stock, the position of the seeds within the capsule, and of the flowers on the spike,—sometimes suffice to determine the variation of a plant, when raised from seed. Now, it is probable, as explained when discussing alternate generation, that a bud is formed of a portion of already differentiated tissue; consequently an organism thus formed does not pass through the earlier phases of development, and cannot be so freely exposed, at the age when its structure would be most readily modified, to the various causes inducing variability; but it is very doubtful whether this is a sufficient explanation of the difficulty.

With respect to the tendency to reversion, there is a similar difference between plants propagated from buds and seed. Many varieties, whether originally produced from seed or buds, can be securely propagated by buds, but generally or invariably revert by seed. So, also, hybridised plants can be multiplied to any extent by buds, but are continually liable to reversion by seed,—that is, to the loss of their hybrid or intermediate character. I can offer no satisfactory explanation of this fact. Here is a still more perplexing case: certain plants with variegated leaves, phloxes with striped flowers, barberries with seedless fruit, can all be securely propagated by the buds on cuttings; but the buds developed from the roots of these cuttings almost invariably lose their character and revert to their former condition.

Finally, we can see on the hypothesis of pangenesis that variability depends on at least two distinct groups of causes. Firstly, on the deficiency, superabundance, fusion, and transposition of gemmules, and on the redevelopment of those which have long been dormant. In these cases the gemmules themselves have undergone no modification; but the mutations in the above respects will amply account for much fluctuating variability. Secondly, in the cases in which the organisation has been modified by changed conditions, the increased use or disuse of parts, or any other cause, the gemmules cast off from the modified units of the body will be themselves modified, and, when sufficiently multiplied, will be developed into new and changed structures.

Turning now to Inheritance; if we suppose a homogeneous gelatinous protozoon to vary and assume a reddish colour, a minute separated atom would naturally, as it grew to full size, retain the same colour; and we should have the simplest form of inheritance. Precisely the same view may be extended to the infinitely numerous and diversified units of which the whole body in one of the higher animals is composed; and the separated atoms are our gemmules. We have already sufficiently discussed the inheritance of the direct effects of changed conditions, and of increased use or disuse of parts, and, by implication, the important principle of inheritance at corresponding ages. These groups of facts are to a large extent intelligible on the hypothesis of pangenesis, and on no other hypothesis as yet advanced.

A few words must be added on the complete abortion or suppression of organs. When a part becomes diminished by disuse prolonged during many generations, the principle of economy of growth, as previously explained, will tend to reduce it still further; but this will not account for the complete or almost complete obliteration of, for instance, a minute papilla of cellular tissue representing a pistil, or of a microscopically minute nodule of bone representing a tooth. In certain cases of suppression not yet completed, in which a rudiment occasionally reappears through reversion, diffused gemmules derived from this part must, according to our view, still exist; hence we must suppose that the cells, in union with which the rudiment was formerly developed, in these cases fail in their affinity for such gemmules. But in the cases of complete and final abortion the gemmules themselves no doubt have perished; nor is this in any way improbable, for, though a vast number of active and long-dormant gemmules are diffused and nourished in each living creature, yet there must be some limit to their number; and it appears natural that gemmules derived from an enfeebled and useless rudiment would be more liable to perish than those derived from other parts which are still in full functional activity.

With respect to mutilations, it is certain that a part may be removed or injured during many generations, and no inherited result follow; and this is an apparent objection to the hypothesis which will occur to every one. But, in the first place, a being can hardly be intentionally mutilated during its early stages of growth whilst in the womb or egg; and such mutilations, when naturally caused, would appear like congenital deficiencies, which are occasionally inherited. In the second place, according to our hypothesis, gemmules multiply by self-division and are transmitted from generation to generation; so that during a long period they would be present and ready to reproduce a part which was repeatedly amputated. Nevertheless it appears, from the facts given in the twelfth chapter, that in some rare cases mutilations have been inherited, but in most of these the mutilated surface became diseased. In this case it may be conjectured that the gemmules of the lost part were gradually all attracted by the partially diseased surface, and thus perished. Although this would occur in the injured individual alone, and therefore in only one parent, yet this might suffice for the inheritance of a mutilation, on the same principle that a hornless animal of either sex, when crossed with a perfect animal of the opposite sex, often transmits its deficiency.

The last subject that need here be discussed, namely Reversion, rests on the principle that transmission and development, though generally acting in conjunction, are distinct powers; and the transmission of gemmules and their subsequent development show us how the existence of these two distinct powers is possible. We plainly see this distinction in the many cases in which a grandfather transmits to his grandson, through his daughter, characters which she does not, or cannot, possess. Why the development of certain characters, not necessarily in any way connected with the reproductive organs, should be confined to one sex alone—that is, why certain cells in one sex should unite with and cause the development of certain gemmules—we do not in the least know; but it is the common attribute of most organic beings in which the sexes are separate.

The distinction between transmission and development is likewise seen in all ordinary cases of Reversion; but before discussing this subject it may be advisable to say a few words on those characters which I have called latent, and which would not be classed under Reversion in its usual sense. Most, or perhaps all, the secondary characters, which appertain to one sex, lie dormant in the other sex; that is, gemmules capable of development into the secondary male sexual characters are included within the female; and conversely female characters are in the male. Why in the female, when her ovaria become diseased or fail to act, certain masculine gemmules become developed, we do not clearly know, any more than why when a young bull is castrated his horns continue growing until they almost resemble those of a cow; or why, when a stag is castrated, the gemmules derived from the antlers of his progenitors quite fail to be developed. But in many cases, with variable organic beings, the mutual affinities of the cells and gemmules become modified, so that parts are transposed or multiplied; and it would appear that a slight change in the constitution of an animal, in connection with the state of the productive organs, leads to changed affinities in the tissues of various parts of the body. Thus, when male animals first arrive at puberty, and subsequently during each recurrent season, certain cells or parts acquire an affinity for certain gemmules, which become developed into the secondary masculine characters; but if the reproductive organs be destroyed, or even temporarily disturbed by changed conditions, these affinities are not excited. Nevertheless, the male, before he arrives at puberty, and during the season when the species does not breed, must include the proper gemmules in a latent state. The curious case formerly given of a Hen which assumed the masculine characters, not of her own breed but of a remote progenitor, illustrates the connection between latent sexual characters and ordinary reversion. With those animals and plants which habitually produce several forms, as with certain butterflies described by Mr. Wallace, in which three female forms and the male exist, or as with the trimorphic species of Lythrum and Oxalis, gemmules capable of reproducing several widely-differently forms must be latent in each individual.

. . .

Reversion, in the ordinary sense of the word, comes into action so incessantly, that it evidently forms an essential part of the general law of inheritance. It occurs with beings, however propagated, whether by buds or seminal generation, and sometimes may even be observed in the same individual as it advances in age. The tendency to reversion is often induced by a change of conditions, and in the plainest manner by the act of crossing. Crossed forms are generally at first nearly intermediate in character between their two parents; but in the next generation the offspring generally revert to one or both of their grandparents, and occasionally to more remote ancestors. How can we account for these facts? Each organic unit in a hybrid must throw off, according to the doctrine of pangenesis, an abundance of hybridised gemmules, for crossed plants can be readily and largely propagated by buds; but by the same hypothesis there will likewise be present dormant gemmules derived from both pure parent-forms; and as these latter retain their normal condition, they would, it is probable, be enabled to multiply largely during the lifetime of each hybrid. Consequently the sexual elements of a hybrid will include both pure and hybridised gemmules; and when two hybrids pair, the combination of pure gemmules derived from the one hybrid with the pure gemmules of the same parts derived from the other would necessarily lead to complete reversion of character; and it is, perhaps, not too bold a supposition that unmodified and undeteriorated gemmules of the same nature would be especially apt to combine. Pure gemmules in combination with hybridised gemmules would lead to partial reversion. And lastly, hybridised gemmules derived

140

from both parent-hybrids would simply reproduce the original hybrid form. All these cases and degrees of reversion incessantly occur.

It was shown in the fifteenth chapter that certain characters are antagonistic to each other or do not really blend together; hence, when two animals with antagonistic characters are crossed, it might well happen that a sufficiency of gemmules in the male alone for the reproduction of his peculiar characters, and in the female alone for the reproduction of her peculiar characters, would not be present; and in this case dormant gemmules derived from some remote progenitor might easily gain the ascendency, and cause the reappearance of long-lost characters. For instance, when black and white pigeons, or black and white fowls, are crossed,—colours which do not readily blend,—blue plumage in the one case, evidently derived from the rock-pigeon, and red plumage in the other case, derived from the wild jungle-cock, occasionally reappear. With uncrossed breeds the same result would follow, under conditions which favoured the multiplication and development of certain dormant gemmules, as when animals become feral and revert to their pristine character. A certain number of gemmules being requisite for the development of each character, as is known to be the case from several spermatozoa or pollen-grains being necessary for fertilisation, and time favouring their multiplication, will together account for the curious cases, insisted on by Mr. Sedgwick, of certain diseases regularly appearing in alternate generations. This likewise holds good, more or less strictly, with other weakly inherited modifications. Hence, as I have heard it remarked, certain diseases appear actually to gain strength by the intermission of a generation. The transmission of dormant gemmules during many successive generations is hardly in itself more improbable, as previously remarked, than the retention during many ages of rudimentary organs, or even only of a tendency to the production of a rudiment; but there is no reason to suppose that all dormant gemmules would be transmitted and propagated for ever. Excessively minute and numerous as they are believed to be, an infinite number derived, during a long course of modification and descent, from each cell of each progenitor, could not be supported or nourished by the organism. On the other hand, it does not seem improbable that certain gemmules, under favourable conditions, should be retained and go on multiplying for a longer period than others. Finally, on the views here given, we certainly gain some clear insight into the wonderful fact that the child may depart from the type of both its parents, and resemble its grandparents, or ancestors removed by many generations.

Conclusion

The hypothesis of Pangenesis, as applied to the several great classes of facts just discussed, no doubt is extremely complex; but so assuredly are the facts. The assumptions, however, on which the hypothesis rests cannot be considered as complex in any extreme degree—namely, that all organic units, besides having the power, as is generally admitted, of growing by self-division, throw off free and minute atoms of their contents, that is gemmules. These multiply and aggregate themselves into buds and the sexual elements; their development depends on their union with other nascent cells or units; and they are capable of transmission in a dormant state to successive generations.

In a highly organised and complex animal, the gemmules thrown off from each different cell or unit throughout the body must be inconceivably numerous and minute. Each unit of each part, as it changes during development, and we know that some insects undergo at least twenty metamorphoses, must throw off its gemmules. All organic beings, moreover, include many dormant gemmules derived from their grandparents and more remote progenitors, but not from all their progenitors. These almost infinitely numerous and minute gemmules must be included in each bud, ovule, spermatozoon, and pollen-grain. Such an admission will be declared impossible; but, as previously remarked, number and size are only relative difficulties, and the eggs or seeds produced by certain animals or plants are so numerous that they cannot be grasped by the intellect.

The organic particles with which the wind is tainted over miles of space by certain offensive animals must be infinitely minute and numerous; yet they strongly affect the olfactory nerves. An analogy more appropriate is afforded by the contagious particles of certain diseases, which are so minute that they float in the atmosphere and adhere to smooth paper; yet we know how largely they increase within the human body, and how powerfully they act. Independent organisms exist which are barely visible under the highest powers of our recently-improved microscopes, and which probably are fully as large as the cells or units in one of the higher animals; yet these organisms no doubt reproduce themselves by germs of extreme minuteness, relatively to their own minute size. Hence the difficulty, which at first appears insurmountable, of believing in the existence of gemmules so numerous and so small as they must be according to our hypothesis, has really little weight.

The cells or units of the body are generally admitted by physiologists to be autonomous, like the buds on a tree, but in a less degree. I go one step further and assume that they throw off reproductive gemmules. Thus an animal does not, as a whole, generate its kind through the sole agency of the reproductive system, but each separate cell generates its kind. It has often been said by naturalists that each cell of a plant has the actual or potential capacity of reproducing the whole plant; but it has this power only in virtue of containing gemmules derived from every part. If our hypothesis be provisionally accepted, we must look at all the forms of asexual reproduction, whether occurring at maturity or as in the case of alternate generation during youth, as fundamentally the same, and dependent on the mutual aggregation and multiplication of the gemmules. The regrowth of an amputated limb or the healing of a wound is the same process partially carried out. Sexual generation differs in some important respects, chiefly, as it would appear, in an insufficient number of gemmules being aggregated within the separate sexual elements, and probably in the presence of certain primordial cells. The development of each being, including all the forms of metamorphosis and metagenesis, as well as the so-called growth of the higher animals, in which structure changes though not in a striking manner, depends on the presence of gemmules thrown off at each period of life, and on their development, at a corresponding period, in union with preceding cells. Such cells may be said to be fertilised by the gemmules which come next in the order of development. Thus the ordinary act of impregnation and the development of each being are closely analogous processes. The child, strictly speaking, does not grow into the man, but includes germs which slowly and successively become developed and form the man. In the child, as well as in the adult, each part generates the same part for the next generation. Inheritance must be looked at as merely a form of growth, like the self-division of a lowly-organised unicellular plant. Reversion depends on the transmission from the forefather to his descendants of dormant gemmules, which occasionally become developed under certain known or unknown conditions. Each animal and plant may be compared to a bed of mould full of seeds, most of which soon germinate, some lie for a period dormant, whilst others perish. When we hear it said that a man carries in his constitution the seeds of an inherited disease, there is much literal truth in the expression. Finally, the power of propagation possessed by each separate cell, using the term in its largest sense, determines the reproduction, the variability, the development and renovation of each living organism. No other attempt, as far as I am aware, has been made, imperfect as this confessedly is, to connect under one point of view these several grand classes of facts. We cannot fathom the marvellous complexity of an organic being; but on the hypothesis here advanced this complexity is much increased. Each living creature must be looked at as a microcosm—a little universe, formed of a host of self-propagating organisms, inconceivably minute and as numerous as the stars in heaven.

Notes

1 Johannes Müller, 1801–1858. German physiologist.

2 Abraham Trembley, 1710–1784. Swiss naturalist; discoverer of *Hydra*.

3 Sir George Edward Paget, 1809–1892. British physician; worked in Cambridge.

4 Sir John Lubbock, 1803–1865. Victorian polymath. Friend of Darwin, and neighbour at Down; MP; writer on innumerable subjects including natural history.

5 Certain fungus feeding species of *Cecidomyiidae* (gall midges, Diptera) can reproduce parthenogenetically as larvae; female oocytes – which Darwin calls germ-balls here – develop within the larvae and give rise to offspring without fertilisation. They can reproduce sexually as adults too.

6 i.e. mayflies.

7 Claude Bernard, 1813–1878. French physiologist. The 'founder of modern experimental physiology', he made important discoveries in digestive, cardiovascular, and nervous physiology.

8 Rudolf Carl Virchow, 1821–1902. 'The most prominent German physician of the nineteenth century.' His most famous work was on pathology and on public health.

9 Lionel Smith Beale, 1828–1906. British micro-anatomist; vitalist; did not accept Darwin's theory of evolution.

10 *Lytta vesicatoria*, the 'Spanish fly', is actually a meloid beetle; it produces the pharmaceutical substance cantharidin, which has been used as an aphrodisiac but really only irritates the urinary tract. The worrara poison is now more often called curare.

11 Lazzaro Spallanzani, 1727–1799. Italian biologist; made many discoveries, of which the most famous is the true function of sperm in sexual reproduction. Charles Bonnet, 1720–1799. Swiss naturalist and philosopher. Preformationist embryologist.

Chapter six

The descent of man, and selection in relation to sex (1871)

The *Descent of man* is a book in three parts, which modern biologists tend to think of as two quite separate books. The first part is on the origin of man; the second – and longest – is on 'sexual selection'; and the third considers how sexual selection may operate in humans. Darwin had been collecting notes on man since 1837, without any particular plan of publication; in 1864 he offered them to Wallace because 'I do not suppose that I shall ever use them.' But Darwin later came to feel that his brief remark about man in the *Origin* – that 'light will be thrown on the origin of man and his history' – was excessively brief. He decided to work up his notes into a book.

The book opens with a discussion of 'the evidence of the descent of man from some lower form', i.e. that man had evolved and not been separately created. The evidence for human evolution is of the same kind as the evidence for evolution in general, which was dealt with in the *Origin*. For some reason, one instance of the argument from rudimentary structures, in this case a detail of the structure of the human ear, particularly captured the Victorian imagination. It was being talked about everywhere. 'The tips of the ears have become quite celebrated', Darwin wrote to Mr Woolner (Figure 6.1). 'A German is very proud to find that he has the tips well developed, and I believe will send me a photograph of his ears.' According to a contemporary journal, 'when Mr Darwin pointed out to us the persistent tips in our ears, he did more to discomfort the friends of persistent species than he did by thousands of other facts.' I could not resist extracting so striking a passage, even though it is only one argument among many.

The process of human evolution is also the same as the general process discussed in the *Origin*. Humans only pose a particular problem in their peculiarly large intelligence, which is of no obvious reproductive advantage, and their moral habits, which seem, if anything, often to reduce rather than increase the number of children left by individual humans. Natural selection ought to work against it. In Chapter 5 of the *Descent*, Darwin considers how natural selection nevertheless might favour moral behaviour in the particular circumstances of human evolution. I have extracted this passage; it is especially interesting because of its relation to the modern theory of altruistic behaviour. Darwin's explanation resembles the modern theories both of 'group' and of 'kin' selection.

Part II of the *Descent* concerns 'selection in relation to sex'. In many species, some individuals – usually males – possess bizarre and exaggerated organs. The peacock's tail is a good example. The large tail blows around in the wind and prevents the bird from flying as well as it otherwise would; and its bright colours will attract enemies: the peacock would probably survive better without its tail. How then could it evolve? Natural selection should act to eliminate it; and yet the peacock does have its tail. Darwin's solution was his theory of sexual selection. He discussed the theory briefly in the *Origin*; but his full argument, and the evidence to support it, came in the *Descent of man*. The peacock's tail could be advantageous if it enabled the peacock to attract more mates than if it lacked so striking and colourful an organ; if an exaggerated tail attracted a sufficiently large number of extra mates it could more than out-weigh the decrease in viability that it caused. That is sexual selection by what Darwin called female choice; he also discussed the process of male competition. Some male organs, unlike the peacock's tail, apparently function in fights among males; the antlers of deer stags are an example: these could also be favoured because they enabled their bearer to obtain more mates, but by fighting off rivals rather than directly attracting females.

I have extracted Darwin's introductory discussion of sexual selection. After the general chapter, Darwin wrote ten chapters on particular groups of animals, to show how important the process is in nature. Sexual selection has favoured modified claws in certain crustacea, the horns of lamellicorn beetles, the sound-producing organs of crickets, the songs and colour patterns of birds, and innumerable other organs. Darwin's is still the most comprehensive review of sexual selection in the whole animal kingdom. He is at his most impressively thorough, as, having worked out a theory of general importance, he arranges an enormous variety of evidence, from all nature, around his theory. For the biologist, that part of the *Descent* on 'selection in relation to sex' is, after the *Origin*, the most important of Darwin's works.

There follow, in order, the sections on the human ear as evidence of our origin 'from a lower form'; the evolution of our moral and intellectual capacities; and finally the main part of the chapter on 'the principles of sexual selection'.

The evidence of the descent of man from some lower form: rudiments

The extrinsic muscles which serve to move the whole external ear, and the intrinsic muscles which move the different parts, all of which belong to the system of the panniculus, are in a rudimentary condition in man; they are also variable in development, or at least in function. I have seen one man who could draw his ears forwards, and another who could draw them backwards; and from what one of these persons told me, it is probable that most of us by often touching our ears and thus directing our attention towards them, could by repeated trials recover some power of movement. The faculty of erecting the ears and of directing them to different points of the compass, is no doubt of the highest service to many animals, as they thus perceive the point of danger; but I have never heard of a man who possessed the least power of erecting his ears,—the one movement which might be of use to him. The whole external shell of the ear may be considered a rudiment, together with the various folds and prominences (helix and anti-helix, tragus and anti-tragus, &c.) which in the lower animals strengthen and support the ear when erect, without adding much to its weight. Some authors, however, suppose that the cartilage of the shell serves to transmit vibrations to the acoustic nerve; but Mr. Toynbee, after collecting all the known evidence on this head, concludes that the external shell is of no distinct use. The ears of the chimpanzee and orang are curiously like those of man, and I am assured by the keepers in the Zoological Gardens that these animals never move or erect them; so that they are in an equally rudimentary condition, as far as function is concerned, as in man. Why these animals, as well as the progenitors of man, should have lost the power of erecting their ears we cannot say. It may be, though I am not quite satisfied with this view, that owing to their arboreal habits and great strength they were but little exposed to danger, and so during a lengthened period moved their ears but little, and thus gradually lost the power of moving them. This would be a parallel case with that of those large and heavy birds, which from inhabiting oceanic islands have not been exposed to the attacks of beasts of prey, and have consequently lost the power of using their wings for flight.

The celebrated sculptor, Mr. Woolner,[1] informs me of one little peculiarity in the external ear, which he has often observed both in men and women, and of which he perceived the full signification. His attention was first called to the subject whilst at work on his figure of Puck, to which he had given pointed ears. He was thus led to examine the ears of various monkeys, and subsequently more carefully those of man. The peculiarity consists in a little blunt point, projecting from the inwardly folded margin, or helix. Mr. Woolner made an exact model of one such case, and has sent me the accompanying drawing [Fig. 6.1]. These points not only project inwards, but often a little outwards, so that they are visible when the head is viewed from directly in front or behind. They are variable in size and somewhat in position, standing either a little higher or lower; and they sometimes occur on one ear and not on the other. Now the meaning of these projections is not, I think, doubtful; but it may be thought that they offer too trifling a character to be worth notice. This thought, however, is as false as it is natural. Every character, however slight, must be the result of some definite cause; and if it occurs in many individuals deserves consideration. The helix obviously consists of the extreme margin of the ear folded inwards; and this folding appears to be in some manner connected with the whole external ear being permanently pressed backwards. In many monkeys, which do

Figure 6.1   The human ear
. . . modelled and drawn by Mr. Woolner. *a*, The projecting point.

not stand high in the order, as baboons and some species of macacus, the upper portion of the ear is slightly pointed, and the margin is not at all folded inwards; but if the margin were to be thus folded, a slight point would necessarily project inwards and probably a little outwards. This could actually be observed in a specimen of the *Ateles beelzebuth* in the Zoological Gardens; and we may safely conclude that it is a similar structure—a vestige of formerly pointed ears—which occasionally reappears in man.

. . .

On the development of the intellectual and moral faculties during primeval and civilised times

The subjects to be discussed in this chapter are of the highest interest, but are treated by me in a most imperfect and fragmentary manner. Mr. Wallace, in an admirable paper . . . argues that man after he had partially acquired those intellectual and moral faculties which distinguish him from the lower animals, would have been but little liable to have had his bodily structure modified through natural selection or any other means. For man is enabled through his mental faculties "to keep with an unchanged body in harmony with the changing universe." He has great power of adapting his habits to new conditions of life. He invents weapons, tools and various stratagems, by which he procures food and defends himself. When he migrates into a colder climate he uses clothes, builds sheds, and makes fires; and, by the aid of fire, cooks food otherwise indigestible. He aids his fellow-men in many ways, and anticipates future events. Even at a remote period he practised some subdivision of labour.

The lower animals, on the other hand, must have their bodily structure modified in order to survive under greatly changed conditions. They must be rendered stronger, or acquire more effective teeth or claws, in order to defend themselves from new enemies; or they must be reduced in size so as to escape detection and danger. When they migrate into a colder climate they must become clothed with thicker fur, or have their constitutions altered. If they fail to be thus modified, they will cease to exist.

The case, however, is widely different, as Mr. Wallace has with justice insisted, in relation to the intellectual and moral faculties of man. These faculties are variable; and we have every reason to believe that the variations tend to be inherited. Therefore, if they were formerly of high importance to primeval man and to his ape-like progenitors, they would have been perfected or advanced through natural selection. Of the high importance of the intellectual faculties there can be no doubt, for man mainly owes to them his pre-eminent position in the world. We can see that, in the rudest state of society, the individuals who were the most sagacious, who invented and used the best weapons or traps, and who were best able to defend themselves, would rear the greatest number of offspring. The tribes which included the largest number of men thus endowed would increase in number and supplant other tribes. Numbers depend primarily on the means of subsistence, and this, partly on the physical nature of the country, but in a much higher degree on the arts which are there practised. As a tribe increases and is victorious, it is often still further increased by the absorption of other tribes. The stature and strength of the men of a tribe are likewise of some importance for its success, and these depend in part on the nature and amount of the food which can be obtained. In Europe the men of the Bronze period were supplanted by a more powerful and, judging from their sword-handles, larger-handed race; but their success was probably due in a much higher degree to their superiority in the arts.

All that we know about savages, or may infer from their traditions and from old monuments, the history of which is quite forgotten by the present inhabitants, shew that from the remotest times successful tribes have supplanted other tribes. Relics of extinct or forgotten tribes have been discovered throughout the civilised regions of the earth, on the wild plains of America, and on the isolated islands in the Pacific Ocean. At the present day civilised nations are everywhere supplanting barbarous nations, excepting where the climate opposes a deadly barrier; and they succeed mainly, though not exclusively, through their arts, which are the products of the intellect. It is, therefore, highly probable that with mankind the intellectual faculties have been gradually perfected through natural selection; and this conclusion is sufficient for our purpose. Undoubtedly it would have been very interesting to have traced the development of each separate faculty from the state in which it exists in the lower animals to that in which it exists in man; but neither my ability nor knowledge permit the attempt.

It deserves notice that as soon as the progenitors of man became social (and this probably occurred at a very early period), the advancement of the intellectual faculties will have been aided and modified in an important manner, of which we see only traces in the lower animals, namely, through the principle of imitation, together with reason and experience. Apes are much given to imitation, as are the lowest savages; and the simple fact previously referred to, that after a time no animal can be caught in the same place by the same sort of trap, shews that animals learn by experience, and imitate each others' caution. Now, if some one man in a tribe, more sagacious than the others, invented a new snare or weapon, or other means of attack or defence, the plainest self-interest, without the assistance of much reasoning power, would prompt the other members to imitate him; and all would thus profit. The habitual practice of each new art must likewise in some slight degree strengthen the intellect. If the new invention were an important one, the tribe would increase in number, spread, and supplant other tribes. In a tribe thus rendered more numerous there would always be a rather better chance of the birth of other superior and inventive members. If such men left children to inherit their mental superiority, the chance of the birth of still more ingenious members would be somewhat better, and in a very small tribe decidedly better. Even if they left no children, the tribe would still include their blood-relations; and it has been ascertained by agriculturists that by preserving and breeding from the family of an animal, which when slaughtered was found to be valuable, the desired character has been obtained.

Turning now to the social and moral faculties. In order that primeval men, or the ape-like progenitors of man, should have become social, they must have acquired the same instinctive feelings which impel other animals to live in a body; and they no doubt exhibited the same general disposition. They would have felt uneasy when separated from their comrades, for whom they would have felt some degree of love; they would have warned each other of danger, and have given mutual aid in attack or defence. All this implies some degree of sympathy, fidelity, and courage. Such social qualities, the paramount importance of which to the lower animals is disputed by no one, were no doubt acquired by the progenitors of man in a similar manner, namely, through natural selection, aided by inherited habit. When two tribes of primeval man, living in the same country, came into competition, if the one tribe included (other circumstances being equal) a greater number of courageous, sympathetic, and faithful members, who were always ready to warn each other of danger, to aid and defend each other, this tribe would without doubt succeed best and conquer the other. Let it be borne in mind how all-important, in the never-ceasing wars of savages, fidelity and courage must be. The advantage which disciplined soldiers have over undisciplined hordes follows chiefly from the confidence which each man feels in his comrades. Obedience, as Mr. Bagehot has well shewn,[2] is of the highest value, for any form of government is better than none. Selfish and contentious people will not cohere, and without coherence nothing can be effected. A tribe possessing the above qualities in a high degree would spread and be victorious over other tribes; but in the course of time it would, judging from all past history, be in its turn overcome by some other and still more highly endowed tribe. Thus the social and moral qualities would tend slowly to advance and be diffused throughout the world.

But it may be asked, how within the limits of the same tribe did a large number of members first become endowed with these social and moral qualities, and how was the standard of excellence raised? It is extremely doubtful whether the offspring of the more sympathetic and benevolent parents, or of those which were the most faithful to their comrades, would be reared in greater number than the children of selfish and treacherous parents of the same tribe. He who was ready to sacrifice his life, as many a savage has been, rather than betray his comrades, would often leave no offspring to inherit his noble nature. The bravest men, who were always willing to come to the front in war, and who freely risked their lives for others, would on an average perish in larger number than other men. Therefore it seems scarcely possible (bearing in mind that we are not here speaking of one tribe being victorious over another) that the number of men gifted with such virtues, or that the standard of their excellence, could be increased through natural selection, that is, by the survival of the fittest.

Although the circumstances which lead to an increase in the number of men thus endowed within the same tribe are too complex to be clearly followed out, we can trace some of the probable steps. In the first place, as the reasoning powers and foresight of the members became improved, each man would soon learn from experience that if he aided his fellow-men, he would commonly receive aid in return. From this low motive he might acquire the habit of aiding his fellows; and the habit of performing benevolent actions certainly strengthens the feeling of sympathy, which gives the first impulse to benevolent actions. Habits, moreover, followed during many generations probably tend to be inherited.

But there is another and much more powerful stimulus to the development of the social virtues, namely, the praise and the blame of our fellow-men. The love of approbation and the dread of infamy, as well as the bestowal of praise or blame, are primarily due, as we have seen in the third chapter, to the instinct of sympathy; and this instinct no doubt was originally acquired, like all the other social instincts, through natural selection. At how early a period the progenitors of man, in the course of their development, became capable of feeling and being impelled by the praise or blame of their fellow-creatures, we cannot, of course, say. But it appears that even dogs appreciate encouragement, praise and blame. The rudest savages feel the sentiment of glory, as they clearly show by preserving the trophies of their prowess, by their habit of excessive boasting, and even by the extreme care which they take of their personal appearance and decorations; for unless they regarded the opinion of their comrades, such habits would be senseless.

They certainly feel shame at the breach of some of their lesser rules; but how far they experience remorse is doubtful. I was at first surprised that I could not recollect any recorded instances of this feeling in savages; and Sir J. Lubbock[3] states that he knows of none. But if we banish from our minds all cases given in novels and plays and in death-bed confessions made to priests, I doubt whether many of us have actually witnessed remorse; though we may have often seen shame and contrition for smaller offences. Remorse is a deeply hidden feeling. It is incredible that a savage, who will sacrifice his life rather than betray his tribe, or one who will deliver himself up as a prisoner rather than break his parole, would not feel remorse in his inmost soul, though he might conceal it, if he had failed in a duty which he held sacred.

We may therefore conclude that primeval man, at a very remote period, would have been influenced by the praise and blame of his fellows. It is obvious, that the members of the same tribe would approve of conduct which appeared to them to be for the general good, and would reprobate that which appeared evil. To do good unto others—to do unto others as ye would they should do unto you,—is the foundation-stone of morality. It is, therefore, hardly possible to exaggerate the importance during rude times of the love of praise and the dread of blame. A man who was not impelled by any deep, instinctive feeling, to sacrifice his life for the good of others, yet was roused to such actions by a sense of glory, would by his example excite the same wish for glory in other men, and would strengthen by exercise the noble feeling of admiration. He might thus do far more good to his tribe than by begetting offspring with a tendency to inherit his own high character.

With increased experience and reason, man perceives the more remote consequences of his actions, and the self-regarding virtues, such as temperance, chastity, &c., which during early times are, as we have before seen, utterly disregarded, come to be highly esteemed or even held sacred. . . . Ultimately a highly complex sentiment, having its first origin in the social instincts, largely guided by the approbation of our fellow-men, ruled by reason, self-interest, and in later times by deep religious feelings, confirmed by instruction and habit, all combined, constitute our moral sense or conscience.

It must not be forgotten that although a high standard of morality gives but a slight or no advantage to each individual man and his children over the other men of the same tribe, yet that an advancement in the standard of morality and an increase in the number of well-endowed men will certainly give an immense advantage to one tribe over another. There can be no doubt that a tribe including many members who, from possessing in a high degree the spirit of patriotism, fidelity, obedience, courage, and sympathy, were always ready to give aid to each other and to sacrifice themselves for the common good, would be victorious over most other tribes; and this would be natural selection. At all times throughout the world tribes have supplanted other tribes; and as morality is one element in their success, the standard of morality and the number of well-endowed men will thus everywhere tend to rise and increase.

It is, however, very difficult to form any judgment why one particular tribe and not another has been successful and has risen in the scale of civilisation. Many savages are in the same condition as when first discovered several centuries ago. As Mr. Bagehot has remarked, we are apt to look at progress as the normal rule in human society; but history refutes this. The ancients did not even entertain the idea; nor do the oriental nations at the present day. According to another high authority, Mr. Maine[4] "the greatest part of mankind has never shewn a particle of desire that its civil institutions should be improved." Progress seems to depend on many concurrent favourable conditions, far too complex to be followed out. But it has often been remarked, that a cool climate from leading to industry and the various arts has been highly favourable, or even indispensable for this end. The Esquimaux, pressed by hard necessity, have succeeded in many ingenious inventions, but their climate has been too severe for continued progress. Nomadic habits, whether over wide plains, or through the dense forests of the tropics, or along the shores of the sea, have in every case been highly detrimental. Whilst observing the barbarous inhabitants of Tierra del Fuego, it struck me that the possession of some property, a fixed abode, and the union of many families under a chief, were the indispensable requisites for civilisation. Such habits almost necessitate the cultivation of the ground; and the first steps in cultivation would probably result . . . from some such accident as the seeds of a fruit-tree falling on a heap of refuse and producing an unusually fine variety. The problem, however, of the first advance of savages towards civilisation is at present much too difficult to be solved.

. . .

Principles of sexual selection

With animals which have their sexes separated, the males necessarily differ from the females in their organs of reproduction; and these afford the primary sexual characters. But the sexes often differ in what Hunter has called secondary sexual characters, which are not directly connected with the act of reproduction; for instance, in the male possessing certain organs of sense or locomotion, of which the female is quite destitute, or in having them more highly-developed, in order that he may readily find or reach her; or again, in the male having special organs of prehension so as to hold her securely. These latter organs of infinitely diversified kinds graduate into, and in some cases can hardly be distinguished from, those which are commonly ranked as primary, such as the complex appendages at the apex of the abdomen in male insects. Unless indeed we confine the term "primary" to the reproductive glands, it is scarcely possible to decide, as far as the organs of prehension are concerned, which ought to be called primary and which secondary.

The female often differs from the male in having organs for the nourishment or protection of her young, as the mammary glands of mammals, and the abdominal sacks of the marsupials. The male, also, in some few cases differs from the female in possessing analogous organs, as the receptacles for the ova possessed by the males of certain fishes, and those temporarily developed in certain male frogs. Female bees have a special apparatus for collecting and carrying pollen, and their ovipositor is modified into a sting for the defence of their larvæ and the community. In the females of many insects the ovipositor is modified in the most complex manner for the safe placing of the eggs. Numerous similar cases could be given, but they do not here concern us. There are, however, other sexual differences quite disconnected with the primary organs with which we are more especially concerned—such as the greater size, strength, and pugnacity of the male, his weapons of offence or means of defence against rivals, his gaudy colouring and various ornaments, his power of song, and other such characters.

Besides the foregoing primary and secondary sexual differences, the male and female sometimes differ in structures connected with different habits of life, and not at all, or only indirectly, related to the reproductive functions. Thus the females of certain flies (Culicidæ and Tabanidæ) are blood-suckers, whilst the males live on flowers and have their mouths destitute of mandibles. The males alone of certain moths and of some crustaceans (*e.g. Tanais*) have imperfect, closed mouths, and cannot feed. The Complemental males of certain cirripedes live like epiphytic plants either on the female or hermaphrodite form, and are destitute of a mouth and prehensile limbs.[5] In these cases it is the male which has been modified and has lost certain important organs, which the females and other members of the same group possess. In other cases it is the female which has lost such parts; for instance, the female glow-worm is destitute of wings, as are many female moths, some of which never leave their cocoons. Many female parasitic crustaceans have lost their natatory legs. In some weevil-beetles (Curculionidæ) there is a great difference between the male and female in the length of the rostrum or snout; but the meaning of this and of many analogous differences, is not at all understood. Differences of structure between the two sexes in relation to different habits of life are generally confined to the lower animals; but with some few birds the beak of the male differs from that of the female. No doubt in most, but apparently not in all these cases, the differences are indirectly connected with the propagation of the species: thus a female which has to nourish a multitude of ova will require more food than the male, and consequently will require special means for procuring it. A male animal which lived for a very short time might without detriment lose through disuse its organs for procuring food; but he would retain his locomotive organs in a perfect state, so that he might reach the female. The female, on the other hand, might safely lose her organs for flying, swimming, or walking, if she gradually acquired habits which rendered such powers useless.

We are, however, here concerned only with that kind of selection, which I have called sexual selection. This depends on the advantage which certain individuals have over other individuals of the same sex and species, in exclusive relation to reproduction. When the two sexes differ in structure in relation to different habits of life, as in the cases above mentioned, they have no doubt been modified through natural selection, accompanied by inheritance limited to one and the same sex. So again the primary sexual organs, and those for nourishing or protecting the young, come under this same head; for those individuals which generated or nourished their offspring best, would leave, *cæteris paribus*, the greatest number to inherit their superiority; whilst those which generated or nourished their offspring badly, would leave but few to inherit their weaker powers. As the male has to search for the female, he requires for this purpose organs of sense and locomotion, but if these organs are necessary for the other purposes of life, as is generally the case, they will have been developed through natural selection. When the male has found the female he sometimes absolutely requires prehensile organs to hold her; thus Dr. Wallace informs me that the males of certain moths cannot unite with the females if their tarsi or feet are broken. The males of many oceanic crustaceans have their legs and antennæ modified in an extraordinary manner for the prehension of the female; hence we may suspect that owing to these animals being washed about by the waves of the open sea, they absolutely require these organs in order to propagate their kind, and if so their development will have been the result of ordinary or natural selection.

When the two sexes follow exactly the same habits of life, and the male has more highly developed sense or locomotive organs than the female, it may be that these in their perfected state are indispensable to the male for finding the female; but in the vast majority of cases, they serve only to give one male an advantage over another, for the less well-endowed males, if time were allowed them, would succeed in pairing with the females; and they would in all other respects, judging from the structure of the female, be equally well adapted for their ordinary habits of life. In such cases sexual selection must have come into action, for the males have acquired their present structure, not from being better fitted to survive in the struggle for existence, but from having gained an advantage over other males[6] and from having transmitted this advantage to their male offspring alone. It was the importance of this distinction which led me to designate this form of selection as sexual selection. So again, if the chief service rendered to the male by his prehensile organs is to prevent the escape of the female before the arrival of other males, or when assaulted by them, these organs will have been perfected through sexual selection, that is by the advantage acquired by certain males over their rivals. But in most cases it is scarcely possible to distinguish between the effects of natural and sexual selection. Whole chapters could easily be filled with details on the differences between the sexes in their sensory, locomotive, and prehensile organs. As, however, these structures are not more interesting than others adapted for the ordinary purposes of life, I shall almost pass them over, giving only a few instances under each class.

There are many other structures and instincts which must have been developed through sexual selection—such as the weapons of offence and the means of defence possessed by the males for fighting with and driving away their rivals—their courage and pugnacity—their ornaments of many kinds—their organs for producing vocal or instrumental music—and their glands for emitting odours; most of these latter structures serving only to allure or excite the female. That these characters are the result of sexual and not of ordinary selection is clear, as unarmed, unornamented, or unattractive males would succeed equally well in the battle for life and in leaving a numerous progeny, if better endowed males were not present. We may infer that this would be the case, for the females, which are unarmed and unornamented, are able to survive and procreate their kind. Secondary sexual characters of the kind just referred to, will be fully discussed in the following chapters, as they are in many respects interesting, but more especially as they depend on the will, choice, and rivalry of the individuals of either sex. When we behold two males fighting for the possession of the female, or several male birds displaying their gorgeous plumage, and performing the strangest antics before an assembled body of females, we cannot doubt that, though led by instinct, they know what they are about, and consciously exert their mental and bodily powers.[7]

In the same manner as man can improve the breed of his game-cocks by the selection of those birds which are victorious in the cockpit, so it appears that the strongest and most vigorous males, or those provided with the best weapons, have prevailed under nature, and have led to the improvement of the natural breed or species. Through repeated deadly contests, a slight degree of variability, if it led to some advantage, however slight, would suffice for the work of sexual selection; and it is certain that secondary sexual characters are eminently variable. In the same manner as man can give beauty, according to his standard of taste, to his male poultry—can give to the Sebright bantam a new and elegant plumage, an erect and peculiar carriage—so it appears that in a state of nature female birds, by having long selected the more attractive males, have added to their beauty. No doubt this implies powers of discrimination and taste on the part of the female which will at first appear extremely improbable; but I hope hereafter to shew that this is not the case.

From our ignorance on several points, the precise manner in which sexual selection acts is to a certain extent uncertain. Nevertheless if those naturalists who already believe in the mutability of species, will read the following chapters,[8] they will, I think, agree with me that sexual selection has played an important part in the history of the

158

organic world. It is certain that with almost all animals there is a struggle between the males for the possession of the female. This fact is so notorious that it would be superfluous to give instances. Hence the females, supposing that their mental capacity sufficed for the exertion of a choice, could select one out of several males. But in numerous cases it appears as if it had been specially arranged that there should be a struggle between many males. Thus with migratory birds, the males generally arrive before the females at their place of breeding, so that many males are ready to contend for each female. The bird-catchers assert that this is invariably the case with the nightingale and blackcap, as I am informed by Mr. Jenner Weir, who confirms the statement with respect to the latter species.

Mr. Swaysland of Brighton, who has been in the habit, during the last forty years, of catching our migratory birds on their first arrival, writes to me that he has never known the females of any species to arrive before their males. During one spring he shot thirty-nine males of Ray's wagtail (*Budytes Raii*) before he saw a single female. Mr. Gould has ascertained by dissection, as he informs me, that male snipes arrive in this country before the females; but this hardly concerns us, as snipes do not breed here. In the case of fish, at the period when the salmon ascend our rivers, the males in large numbers are ready to breed before the females. So it apparently is with frogs and toads. Throughout the great class of insects the males almost always emerge from the pupal state before the other sex, so that they generally swarm for a time before any females can be seen. The cause of this difference between the males and females in their periods of arrival and maturity is sufficiently obvious. Those males which annually first migrated into any country, or which in the spring were first ready to breed, or were the most eager, would leave the largest number of offspring; and these would tend to inherit similar instincts and constitutions. On the whole there can be no doubt that with almost all animals, in which the sexes are separate, there is a constantly recurrent struggle between the males for the possession of the females.

Our difficulty in regard to sexual selection lies in understanding how it is that the males which conquer other males, or those which prove the most attractive to the females, leave a greater number of offspring to inherit their superiority than the beaten and less attractive males. Unless this result should follow, the characters which give to certain males an advantage over others, could not be perfected and augmented through sexual selection. When the sexes exist in exactly equal numbers, the worst-endowed males will ultimately find females (except where polygamy prevails), and leave as many offspring, equally well fitted for their general habits of life, as the best-endowed males. From various facts and considerations, I formerly inferred that with most animals, in which secondary sexual characters are well developed, the males considerably exceeded the females in number; and this does hold good in some few cases. If the males were to the females as two to one, or as three to two, or even in a somewhat lower ratio, the whole affair would be simple; for the better-armed or more attractive males would leave the largest number of offspring. But after investigating, as far as possible, the numerical proportions of the sexes, I do not believe that any great inequality in number commonly exists. In most cases sexual selection appears to have been effective in the following manner.

Let us take any species, a bird for instance,[9] and divide the females inhabiting a district into two equal bodies: the one consisting of the more vigorous and better-nourished individuals, and the other of the less vigorous and healthy. The former, there can be little doubt, would be ready to breed in the spring before the others; and this is the opinion of Mr. Jenner Weir, who has during many years carefully attended to the habits of birds. There can also be no doubt that the most vigorous, healthy, and best-nourished females would on an average succeed in rearing the largest number of offspring. The males, as we have seen, are generally ready to breed before the females; of the males the strongest, and with some species the best armed, drive away the weaker males; and the former would then unite with the more vigorous and best-nourished females, as these are the first to breed. Such vigorous pairs would surely rear a larger number of offspring than the retarded females, which would be compelled, supposing the sexes to be numerically equal, to unite with the conquered and less powerful males; and this is all that is wanted to add, in the course of successive generations, to the size, strength and courage of the males, or to improve their weapons.

But in a multitude of cases the males which conquer other males, do not obtain possession of the females, independently of choice on the part of the latter. The courtship of animals is by no means so simple and short an affair as might be thought. The females are most excited by, or prefer pairing with, the more ornamented males, or those which are the best songsters, or play the best antics; but it is obviously probable, as has been actually observed in some cases, that they would at the same time prefer the more vigorous and lively males. Thus the more vigorous females, which are the first to breed, will have the choice of many males; and though they may not always select the strongest or best armed, they will select those which are vigorous and well armed, and in other respects the most attractive. Such early pairs would have the same advantage in rearing offspring on the female side as above explained, and nearly the same advantage on the male side. And this apparently has sufficed during a long course of generations to add not only to the strength and fighting-powers of the males, but likewise to their various ornaments or other attractions.

. . .

Numerical Proportion of the Two Sexes.—I have remarked that sexual selection would be a simple affair if the males considerably exceeded in number the females. Hence I was led to investigate, as far as I could, the proportions between the two sexes of as many animals as possible; but the materials are scanty. I will here give only a brief abstract of the results. . . . Domesticated animals alone afford the opportunity of ascertaining the proportional numbers at birth; but no records have been specially kept for this purpose. By indirect means, however, I have collected a considerable body of statistical data, from which it appears that with most of our domestic animals the sexes are nearly equal at birth. Thus with race-horses, 25,560 births have been recorded during twenty-one years, and the male births have been to the female births as 99·7 to 100. With greyhounds the inequality is greater than with any other animal, for during twelve years, out of 6878 births, the male births have been as 110·1 to 100 female births. It is, however, in some degree doubtful whether it is safe to infer that the same proportional numbers would hold good under natural conditions as under domestication; for slight and unknown differences in the conditions affect to a certain extent the proportion of the sexes. Thus with mankind, the male births in England are as 104·5, in Russia as 108·9, and with the Jews of Livonia as 120 to 100 females. The proportion is also mysteriously affected by the circumstance of the births being legitimate or illegitimate.

For our present purpose we are concerned with the proportion of the sexes, not at birth, but at maturity, and this adds another element of doubt; for it is a well ascertained fact that with man a considerably larger proportion of males than of females die before or during birth, and during the first few years of infancy. So it almost certainly is with male lambs, and so it may be with the males of other animals. The males of some animals kill each other by fighting; or they drive each other about until they become greatly emaciated. They must, also, whilst wandering about in eager search for the females, be often exposed to various dangers. With many kinds of fish the males are much smaller than the females, and they are believed often to be devoured by the latter or by other fishes. With some birds the females appear to die in larger proportion than the males: they are also liable to be destroyed on their nests, or whilst in charge of their young.

With insects the female larvæ are often larger than those of the males, and would consequently be more likely to be devoured: in some cases the mature females are less active and less rapid in their movements than the males, and would not be so well able to escape from danger. Hence, with animals in a state of nature, in order to judge of the proportions of the sexes at maturity, we must rely on mere estimation; and this, except perhaps when the inequality is strongly marked, is but little trustworthy. Nevertheless, as far as a judgment can be formed, we may conclude from the facts given in the supplement, that the males of some few mammals, of many birds, of some fish and insects, considerably exceed in number the females.

The proportion between the sexes fluctuates slightly during successive years: thus with race-horses, for every 100 females born, the males varied from 107·1 in one year to 92·6 in another year, and with greyhounds from 116·3 to 95·3. But had larger numbers been tabulated throughout a more extensive area than England, these fluctuations would probably have disappeared; and such as they are, they would hardly suffice to lead under a state of nature to the effective action of sexual selection. Nevertheless with some few wild animals, the proportions seem, as shewn in the supplement, to fluctuate either during different seasons or in different localities in a sufficient degree to lead to such action. For it should be observed that any advantage gained during certain years or in certain localities by those males which were able to conquer other males, or were the most attractive to the females, would probably be transmitted to the offspring and would not subsequently be eliminated. During the succeeding seasons, when from the equality of the sexes every male was everywhere able to procure a female, the stronger or more attractive males previously produced would still have at least as good a chance of leaving offspring as the less strong or less attractive.

Polygamy.—The practice of polygamy leads to the same results as would follow from an actual inequality in the number of the sexes; for if each male secures two or more females, many males will not be able to pair; and the latter assuredly will be the weaker or less attractive individuals. Many mammals and some few birds are polygamous, but with animals belonging to the lower classes I have found no evidence of this habit. The intellectual powers of such animals are, perhaps, not sufficient to lead them to collect and guard a harem of females. That some relation exists between polygamy and the development of secondary sexual characters, appears nearly certain; and this supports the view that a numerical preponderance of males would be eminently favourable to the action of sexual selection. Nevertheless many animals, especially birds, which are strictly monogamous, display strongly-marked secondary sexual characters; whilst some few animals, which are polygamous, are not thus characterised.[10]

We will first briefly run through the class of mammals, and then turn to birds. The gorilla seems to be a polygamist, and the male differs considerably from the female; so it is with some baboons which live in herds containing twice as many adult females as males. In South America the *Mycetes caraya* presents well-marked sexual differences in colour, beard, and vocal organs, and the male generally lives with two or three wives: the male of the *Cebus capucinus* differs somewhat from the female, and appears to be polygamous. Little is known on this head with respect to most other monkeys, but some species are strictly monogamous. The ruminants are eminently polygamous, and they more frequently present sexual differences than almost any other group of mammals, especially in their weapons, but likewise in other characters. Most deer, cattle, and sheep are polygamous; as are most antelopes, though some of the latter are monogamous. Sir Andrew Smith, in speaking of the antelopes of South Africa, says that in herds of about a dozen there was rarely more than one mature male. The Asiatic *Antilope saiga* appears to be the most inordinate polygamist in the world; for Pallas states that the male drives away all rivals, and collects a herd of about a hundred, consisting of females and kids: the female is hornless and has softer hair, but does not otherwise differ much from the male. The horse is polygamous, but, except in his greater size and in the proportions of his body, differs but little from the mare. The wild boar, in his great tusks and some other characters, presents well-marked sexual characters; in Europe and in India he leads a solitary life, except during the breeding-season; but at this season he consorts in India with several females, as Sir W. Elliot, who has had large experience in observing this animal, believes: whether this holds good in Europe is doubtful, but is supported by some statements. The adult male Indian elephant, like the boar, passes much of his time in solitude; but when associating with others, "it is rare to find," as Dr. Campbell states, "more than one male with a whole herd of females." The larger males expel or kill the smaller and weaker ones. The male differs from the female by his immense tusks and greater size, strength, and endurance; so great is the difference in these latter respects, that the males when caught are valued at twenty per cent. above the females. With other pachydermatous animals the sexes differ very little or not at all, and they are not, as far as known, polygamists. Hardly a single species amongst the Cheiroptera and Edentata, or in the great Orders of the Rodents and Insectivora, presents well-developed secondary sexual differences; and I can find no account of any species being polygamous, excepting, perhaps, the common rat, the males of which, as some rat-catchers affirm, live with several females.

The lion in South Africa, as I hear from Sir Andrew Smith, sometimes lives with a single female, but generally with more than one, and, in one case, was found with as many as five females, so that he is polygamous. He is, as far as I can discover, the sole polygamist in the whole group of the terrestrial Carnivora, and he alone presents well-marked sexual characters. If, however, we turn to the marine Carnivora, the case is widely different; for many species of seals offer, as we shall hereafter see, extraordinary sexual differences, and they are eminently polygamous. Thus the male sea-elephant of the Southern Ocean, always possesses, according to Péron, several females, and the sea-lion of Forster is said to be surrounded by from twenty to thirty females. In the North, the male sea-bear of Steller is accompanied by even a greater number of females.

With respect to birds, many species, the sexes of which differ greatly from each other, are certainly monogamous. In Great Britain we see well-marked sexual differences in, for instance, the wild-duck which pairs with a single female, with the common blackbird, and with the bullfinch which is said to pair for life. So it is, as I am informed by Mr. Wallace, with the Chatterers or Cotingidæ of South America, and numerous other birds. In several groups I have not been able to discover whether the species are polygamous or monogamous. Lesson says that birds of paradise, so remarkable for their sexual differences, are polygamous, but Mr. Wallace doubts whether he had sufficient evidence. Mr. Salvin informs me that he has been led to believe that humming-birds are polygamous. The male widow-bird, remarkable for his caudal plumes, certainly seems to be a polygamist. I have been assured by Mr. Jenner Weir and by others, that three starlings not rarely frequent the same nest; but whether this is a case of polygamy or polyandry has not been ascertained.

The Gallinaceæ present almost as strongly marked sexual differences as birds of paradise or humming-birds, and many of the species are, as is well known, polygamous; others being strictly monogamous. What a contrast is presented between the sexes of the polygamous peacock or pheasant, and the monogamous guinea-fowl or partridge! Many similar cases could be given, as in the grouse tribe, in which the males of the polygamous capercailzie and black-cock differ greatly from the females, whilst the sexes of the monogamous red grouse and ptarmigan differ very little. Amongst the Cursores, no great number of species offer strongly-marked sexual differences, except the bustards, and the great bustard (*Otis tarda*), is said to be polygamous. With the Grallatores, extremely few species differ sexually, but the ruff (*Machetes pugnax*) affords a strong exception, and this species is believed by Montagu to be a polygamist. Hence it appears that with birds there often exists a close relation between polygamy and the development of strongly-marked sexual differences. On asking Mr. Bartlett, at the Zoological Gardens, who has had such large experience with birds, whether the male tragopan (one of the Gallinaceæ) was polygamous, I was struck by his answering, "I do not know, but should think so from his splendid colours."

It deserves notice that the instinct of pairing with a single female is easily lost under domestication. The wild-duck is strictly monogamous, the domestic-duck highly polygamous. The Rev. W. D. Fox informs me that with some half-tamed wild-ducks, kept on a large pond in his neighbourhood, so many mallards were shot by the gamekeeper that only one was left for every seven or eight females; yet unusually large broods were reared. The guinea-fowl is strictly monogamous; but Mr. Fox finds that his birds succeed best when he keeps one cock to two or three hens. Canary-birds pair in a state of nature, but the breeders in England successfully put one male to four or five females, nevertheless the first female, as Mr. Fox has been assured, is alone treated as the wife, she and her young ones being fed by him; the others are treated as concubines. I have noticed these cases, as it renders it in some degree probable that monogamous species, in a state of nature, might readily become either temporarily or permanently polygamous.

With respect to reptiles and fishes, too little is known of their habits to enable us to speak of their marriage arrangements. The stickle-back (*Gasterosteus*), however, is said to be a polygamist; and the male during the breeding-season differs conspicuously from the female.

To sum up on the means through which, as far as we can judge, sexual selection has led to the development of secondary sexual characters. It has been shewn that the largest number of vigorous offspring will be reared from the pairing of the strongest and best-armed males, which have conquered other males, with the most vigorous and best-nourished females, which are the first to breed in the spring. Such females, if they select the more attractive, and at the same time vigorous, males, will rear a larger number of offspring than the retarded females, which must pair with the less vigorous and less attractive males. So it will be if the more vigorous males select the more attractive and at the same time healthy and vigorous females; and this will especially hold good if the male defends the female, and aids in providing food for the young. The advantage thus gained by the more vigorous pairs in rearing a larger number of offspring has apparently sufficed to render sexual selection efficient. But a large preponderance in number of the males over the females would be still more efficient; whether the preponderance was only occasional and local, or permanent; whether it occurred at birth, or subsequently from the greater destruction of the females; or whether it indirectly followed from the practice of polygamy.

The Male generally more modified than the Female.—Throughout the animal kingdom, when the sexes differ from each other in external appearance, it is the male which, with rare exceptions, has been chiefly modified; for the female still remains more like the young of her own species, and more like the other members of the same group. The cause of this seems to lie in the males of almost all animals having stronger passions than the females. Hence it is the males that fight together and sedulously display their charms before the females; and those which are victorious transmit their superiority to their male offspring.... That the males of all mammals eagerly pursue the females is notorious to every one. So it is with birds; but many male birds do not so much pursue the female, as display their plumage, perform strange antics, and pour forth their song, in her presence. With the few fish which have been observed, the male seems much more eager than the female; and so it is with alligators, and apparently with Batrachians. Throughout the enormous class of insects, as Kirby remarks, "the law is, that the male shall seek the female." With spiders and crustaceans, as I hear from two great authorities, Mr. Blackwall and Mr. C. Spence Bate, the males are more active and more erratic in their habits than the females. With insects and crustaceans, when the organs of sense or locomotion are present in the one sex and absent in the other, or when, as is frequently the case, they are more highly developed in the one than the other, it is almost invariably the male, as far as I can discover, which retains such organs, or has them most developed; and this shews that the male is the more active member in the courtship of the sexes.

The female, on the other hand, with the rarest exception, is less eager than the male. As the illustrious Hunter long ago observed, she generally "requires to be courted"; she is coy, and may often be seen endeavouring for a long time to escape from the male. Every one who has attended to the habits of animals will be able to call to mind instances of this kind. Judging from various facts, hereafter to be given, and from the results which may fairly be attributed to sexual selection, the female, though comparatively passive, generally exerts some choice and accepts one male in preference to others. Or she may accept, as appearances would sometimes lead us to believe, not the male which is the most attractive to her, but the one which is the least distasteful. The exertion of some choice on the part of the female seems almost as general a law as the eagerness of the male.

We are naturally led to enquire why the male in so many and such widely distinct classes has been rendered more eager than the female, so that he searches for her and plays the more active part in courtship. It would be no advantage and some loss of power if both sexes were mutually to search for each other; but why should the male almost always be the seeker? With plants, the ovules after fertilisation have to be nourished for a time; hence the pollen is necessarily brought to the female organs—being placed on the stigma, through the agency of insects or of the wind, or by the spontaneous movements of the stamens; and with the Algæ, &c., by the locomotive power of the antherozooids. With lowly-organised animals permanently affixed to the same spot and having their sexes separate, the male element is invariably brought to the female; and we can see the reason why; for the ova, even if detached before being fertilised and not requiring subsequent nourishment or protection, would be, from their larger relative size, less easily transported than the male element. Hence plants and many of the lower animals are, in this respect, analogous. The males of affixed animals having been thus led to emit their fertilising element, it is natural that any of their descendants, which rose in the scale and became locomotive, should retain the same habit, and should closely approach the female, so that the fertilising element might not run the risk of a long transit through the waters of the sea. With some few of the lower animals, the females alone are fixed, and with these the males must be the seekers. With respect to forms, of which the progenitors were primordially free, it is difficult to understand why the males should invariably have acquired the habit of approaching the females, instead of being approached by them. But in all cases, in order that the males should be efficient seekers, it would be necessary that they should be endowed with strong passions; and the acquirement of such passions would naturally follow from the more eager males leaving a larger number of offspring than the less eager.

. . .

In various classes of animals a few exceptional cases occur, in which the female instead of the male has acquired well pronounced secondary sexual characters, such as brighter colours, greater size, strength, or pugnacity. With birds, . . . there has sometimes been a complete transposition of the ordinary characters proper to each sex; the females having become the more eager in courtship, the males remaining comparatively passive, but apparently selecting, as we may infer from the results, the more attractive females. Certain female birds have thus been rendered more highly coloured or otherwise ornamented, as well as more powerful and pugnacious than the males, these characters being transmitted to the female offspring alone.[11]

It may be suggested that in some cases a double process of selection has been carried on; the males having selected the more attractive females, and the latter the more attractive males. This process however, though it may lead to the modification of both sexes, would not make the one sex different from the other, unless indeed their taste for the beautiful differed; but this is a supposition too improbable in the case of any animal, excepting man, to be worth considering. There are, however, many animals, in which the sexes resemble each other, both being furnished with the same ornaments, which analogy would lead us to attribute to the agency of sexual selection. In such cases it may be suggested with more plausibility, that there has been a double or mutual process of sexual selection; the more vigorous and precocious females having selected the more attractive and vigorous males, the latter having rejected all except the more attractive females. But from what we know of the habits of animals, this view is hardly probable, the male being generally eager to pair with any female. It is more probable that the ornaments common to both sexes were acquired by one sex, generally the male, and then transmitted to the offspring of both sexes. If, indeed, during a lengthened period the males of any species were greatly to exceed the females in number, and then during another lengthened period under different conditions the reverse were to occur, a double, but not simultaneous, process of sexual selection might easily be carried on, by which the two sexes might be rendered widely different.

. . . Many animals exist of which neither sex is brilliantly coloured or provided with special ornaments, and yet the members of both sexes or of one alone have probably been modified through sexual selection. The absence of bright tints or other ornaments may be the result of variations of the right kind never having occurred, or of the animals themselves preferring simple colours, such as plain black or white. Obscure colours have often been acquired through natural selection for the sake of protection, and the acquirement through sexual selection of conspicuous colours, may have been checked from the danger thus incurred. But in other cases the males have probably struggled together during long ages, through brute force, or by the display of their charms, or by both means combined, and yet no effect will have been produced unless a larger number of offspring were left by the more successful males to inherit their superiority, than by the less successful males; and this, as previously shewn, depends on various complex contingencies.

Sexual selection acts in a less rigorous manner than natural selection. The latter produces its effects by the life or death of all ages of the more or less successful individuals. Death, indeed, not rarely ensues from the conflicts of rival males. But generally the less successful male merely fails to obtain a female, or obtains a retarded and less vigorous female later in the season, or, if polygamous, obtains fewer females; so that they leave fewer, or less vigorous, or no offspring. In regard to structures acquired through ordinary or natural selection, there is in most cases, as long as the conditions of life remain the same, a limit to the amount of advantageous modification in relation to certain special ends; but in regard to structures adapted to make one male victorious over another, either in fighting or in charming the female, there is no definite limit to the amount of advantageous modification; so that as long as the proper variations arise the work of sexual selection will go on. This circumstance may partly account for the frequent and extraordinary amount of variability presented by secondary sexual characters. Nevertheless, natural selection will determine that characters of this kind shall not be acquired by the victorious males, which would be injurious to them in any high degree, either by expending too much of their vital powers, or by exposing them to any great danger. The development, however, of certain structures—of the horns, for instance, in certain stags—has been carried to a wonderful extreme; and in some instances to an extreme which, as far as the general conditions of life are concerned, must be slightly injurious to the male. From this fact we learn that the advantages which favoured males have derived from conquering other males in battle or courtship, and thus leaving a numerous progeny, have been in the long run greater than those derived from rather more perfect adaptation to the external conditions of life.

Notes

1   Thomas Woolner, 1825–1892. Poet and sculptor. Member of the Pre-Raphaelite Brotherhood.

2   W. Bagehot (pronounce *Badg*et), 1826–1877. Writer on politics and economics. Author of *The English constitution*, and of *Physics and politics* to which Darwin refers here.

3   Sir John Lubbock. See Ch. 5, n4. Darwin here refers to Lubbock's *Origin of civilization* (1870).

4   Sir Henry Maine, 1822–1888. A founder of Victorian anthropology. Darwin here refers to his *Ancient law* (1871).

5   Darwin himself had discovered these remarkable tiny males, which live attached to the females in some species of barnacles (*Cirripedia*).

6   N.B. This is Darwin's definition of sexual selection. We are to distinguish organs making males 'better fitted to survive' (which are due to natural selection) from those giving 'an advantage over other males' (due to sexual selection).

7   Obviously a controversial remark. However, the truth of his theory of sexual selection does not depend on it.

8   Not included here: the chapters compile evidence of sexual differences in many species of animals.

9   The following idea is particularly applicable to birds. Darwin considers a species in which the male and female pair for some time and rear their young together; this arrangement is exceptional in the animal kingdom as a whole, but common in birds.

10   Darwin now tests the idea by seeking his predicted association between polygamy and well-marked sexual differences, and monogamy and greater similarity of the sexes.

11   Phalaropes are an example.

Chapter seven

The expression of the emotions in man and animals (1872)

The *Expression of the emotions* formed part of Darwin's larger investigation of man. Indeed, when Darwin first planned the *Descent of man* he thought he would cover the emotions in one chapter; but 'as soon as I began to put my notes together, I saw that it would require a separate Treatise'. He started work on the book immediately he had finished the *Descent*: the last proofs of the *Descent* were passed on 15 January 1871; on the 17th he began the *Expression of the emotions*, and had written the first draft by April; in December the proofs arrived. Darwin was then utterly exhausted. 'I am growing old and weak,' he told Haeckel, 'and no man can tell when his intellectual powers begin to fail . . . I have resumed some old botanical work, and perhaps I shall never again attempt to discuss theoretical views.'

In the *Expression of the emotions* and in the *Descent of man* Darwin's method was the same. He read widely and collected notes on anything that (without any exact aim in mind) might turn out to be useful; he asked his numerous correspondents about points of detail; he drew on his own casual observations; and he thought through the whole subject in depth, and with such originality that he would set it on a wholly new, and solid, base. For his work on the emotions he made relatively more use of casual observation and correspondence, especially with 'missionaries and others living among savages'; but the general method was the same. His work on the subject was 'commenced in the year 1838'; he was then further inspired by the birth of his first child in December 1839 – his children afforded abundant opportunities for observation – and a reading of Sir Charles Bell's *The anatomy and philosophy of expression*. According to Bell, certain muscles had been created in humans specially for the purpose of emotional expression, which clearly contradicted Darwin's theory; all parts of all species should have precursors in related species. So Darwin could write to Wallace (in 1867), 'I want to upset Sir C. Bell's view.' But he would upset it comprehensively, by setting up an entirely new, evolutionary theory; and the original inspiration can hardly be detected in the finished book. In fact it draws on Bell for many particular ideas, when they do not contradict Darwin's evolutionary system.

And what was Darwin's system? He explained the expression of the emotions by three principles, which he called 'serviceable associated habits', 'the principle of antithesis', and 'direct action of the nervous system'. A serviceable associated habit is a habit associated with the emotion being expressed, which is also adaptive; hitting, for example, is a habit serviceably associated with the emotion of anger, and tensing the muscles, as if preparatory to strike, is therefore a serviceable associated habit which might express anger. By the principle of antithesis, an opposite emotion may be expressed by an opposite habit: if we are feeling deferential, we may relax our muscles and generally adopt the posture opposite to that we should take up if we were angry with some one. The third principle, the direct action of the nervous system, Darwin formulated less clearly, but seemed to be required to account for some expressions not easily explained by the other two principles. Trembling, when we feel fear, was Darwin's clearest example; but Darwin was aware that the principle was unsatisfactory. As he wrote to Prof. Alexander Bain (who had criticised it) 'what you say about the vagueness of what

I have called the direct action of the nervous system, is perfectly just. I felt it so at the time, and even more of late.'

The *Expression of the emotions* first explains these three principles; then follow two chapters on emotional expression in animals, before the main material of the book. Seven chapters deal in turn with the various emotions in man: 'suffering and weeping', and 'surprise', for example; the penultimate chapter on blushing is perhaps the best known. Darwin uses his three principles to explain the form of our emotional expression in each case; he makes most use of the principle of serviceable association. I have extracted Darwin's account of the three principles, together with two particular applications: 'indignation' to illustrate a serviceable associated habit and 'shrugging of shoulders' to illustrate antithesis.

The book has influenced later biological thought about communication between animals. The principle of 'serviceable associated habits' can help to explain the form of certain animal signals, particularly those with their origin in what are now called intention movements. The influence has been quite small, however, because the book was concerned more with the expression than with the communication of emotions; and it has more to say about humans than about other animals.

It has continued to be read by an audience beyond biology, for it is not difficult to follow, has a pleasantly anecdotal style, and reflects on everyday observations. Its illustrations too are of some historical interest. Darwin collaborated with an influential and important photographer, Otto Rejlander (a Swede, but naturalised in England). 'All these photographs' Darwin assured the reader, 'have been printed by the Heliotype process, and the accuracy of the copy is thus guaranteed.'

General principles of expression

I will begin by giving the three Principles, which appear to me to account for most of the expressions and gestures involuntarily used by man and the lower animals, under the influence of various emotions and sensations. . . . I need hardly premise that movements or changes in any part of the body,—as the wagging of a dog's tail, the drawing back of a horse's ears, the shrugging of a man's shoulders, or the dilatation of the capillary vessels of the skin,—may all equally well serve for expression. The three Principles are as follows.

I. *The principle of serviceable associated Habits.*—Certain complex actions are of direct or indirect service under certain states of the mind, in order to relieve or gratify certain sensations, desires, &c.; and whenever the same state of mind is induced, however feebly, there is a tendency through the force of habit and association for the same movements to be performed, though they may not then be of the least use. Some actions ordinarily associated through habit with certain states of the mind may be partially repressed through the will, and in such cases the muscles which are least under the separate control of the will are the most liable still to act, causing movements which we recognise as expressive. In certain other cases the checking of one habitual movement requires other slight movements; and these are likewise expressive.

II. *The principle of Antithesis.*—Certain states of the mind lead to certain habitual actions, which are of service, as under our first principle. Now when a directly opposite state of mind is induced, there is a strong and involuntary tendency to the performance of movements of a directly opposite nature, though these are of no use; and such movements are in some cases highly expressive.

III. *The principle of actions due to the constitution of the Nervous System, independently from the first of the Will, and independently to a certain extent of Habit.*—When the sensorium is strongly excited, nerve-force is generated in excess, and is transmitted in certain definite directions, depending on the connection of the nerve-cells, and partly on habit: or the supply of nerve-force may, as it appears, be interrupted. Effects are thus produced which we recognise as expressive. This third principle may, for the sake of brevity, be called that of the direct action of the nervous system.

The principle of serviceable associated habits

With respect to our *first Principle*, it is notorious how powerful is the force of habit. The most complex and difficult movements can in time be performed without the least effort or consciousness. It is not positively known how it comes that habit is so efficient in facilitating complex movements; but physiologists admit "that the conducting power of the nervous fibres increases with the frequency of their excitement." This applies to the nerves of motion and sensation, as well as to those connected with the act of thinking. That some physical change is produced in the nerve-cells or nerves which are habitually used can hardly be doubted, for otherwise it is impossible to understand how the tendency to certain acquired movements is inherited. That they are inherited we see with horses in certain transmitted paces, such as cantering and ambling, which are not natural to them,—in the pointing of young pointers and the setting of young setters—in the peculiar manner of flight of certain breeds of the pigeon, &c. We have analogous cases with mankind in the inheritance of tricks or unusual gestures, to which we shall presently recur. To those who admit the gradual evolution of species, a most striking instance of the perfection with which the most difficult consensual movements can be transmitted, is afforded by the humming-bird Sphinx-moth (*Macroglossa*); for this moth, shortly after its emergence from the cocoon, as shown by the bloom on its unruffled scales, may be seen poised stationary in the air, with its long hair-like proboscis uncurled and inserted into the minute orifices of flowers; and no one, I believe, has ever seen this moth learning to perform its difficult task, which requires such unerring aim.

When there exists an inherited or instinctive tendency to the performance of an action, or an inherited taste for certain kinds of food, some degree of habit in the individual is often or generally requisite. We find this in the paces of the horse, and to a certain extent in the pointing of dogs; although some young dogs point excellently the first time they are taken out, yet they often associate the proper inherited attitude with a wrong odour, and even with eyesight. I have heard it asserted that if a calf be allowed to suck its mother only once, it is much more difficult afterwards to rear it by hand. Caterpillars which have been fed on the leaves of one kind of tree, have been known to perish from hunger rather than to eat the leaves of another tree, although this afforded them their proper food, under a state of nature, and so it is in many other cases.

The power of Association is admitted by everyone. Mr. Bain remarks, that "actions, sensations, and states of feeling, occurring together or in close succession, tend to grow together, or cohere, in such a way that when any one of them is afterwards presented to the mind, the others are apt to be brought up in idea." It is so important for our purpose fully to recognise that actions readily become associated with other actions and with various states of the mind, that I will give a good many instances, in the first place relating to man, and afterwards to the lower animals. Some of the instances are of a very trifling nature, but they are as good for our purpose as more important habits. It is known to everyone how difficult, or even impossible it is, without repeated trials, to move the limbs in certain opposed directions which have never been practised. Analogous cases occur with sensations, as in the common experiment of rolling a marble beneath the tips of two crossed fingers, when it feels exactly like two marbles. Everyone protects himself when falling to the ground by extending his arms, and as Professor Alison has remarked, few can resist acting thus, when voluntarily falling on a soft bed. A man when going out of doors puts on his gloves quite unconsciously; and this may seem an extremely simple operation, but he who has taught a child to put on gloves, knows that this is by no means the case.

When our minds are much affected, so are the movements of our bodies; but here another principle besides habit, namely the undirected overflow of nerve-force, partially comes into play. Norfolk, in speaking of Cardinal Wolsey, says—

> "Some strange commotion
> Is in his brain; he bites his lip and starts;
> Stops on a sudden, looks upon the ground,
> Then, lays his finger on his temple: straight,
> Springs out into fast gait; then, stops again,
> Strikes his breast hard; and anon, he casts
> His eye against the moon: in most strange postures
> We have seen him set himself."

—Hen. VIII., act 3, sc. 2.

A vulgar man often scratches his head when perplexed in mind; and I believe that he acts thus from habit, as if he experienced a slightly uncomfortable bodily sensation, namely, the itching of his head, to which he is particularly liable, and which he thus relieves. Another man rubs his eyes when perplexed, or gives a little cough when embarrassed, acting in either case as if he felt a slightly uncomfortable sensation in his eyes or windpipe.

From the continued use of the eyes, these organs are especially liable to be acted on through association under various states of the mind, although there is manifestly nothing to be seen. A man, as Gratiolet remarks, who vehemently rejects a proposition, will almost certainly shut his eyes or turn away his face; but if he accepts the proposition, he will nod his head in affirmation and open his eyes widely. The man acts in this latter case as if he clearly saw the thing, and in the former case as if he did not or would not see it. I have noticed that persons in describing a horrid sight often shut their eyes momentarily and firmly, or shake their heads, as if not to see or to drive away something disagreeable; and I have caught myself, when thinking in the dark of a horrid spectacle, closing my eyes firmly. In looking suddenly at any object, or in looking all around, everyone raises his eyebrows, so that the eyes may be quickly and widely opened; and Duchenne remarks that a person in trying to remember

176

something often raises his eyebrows, as if to see it. A Hindoo gentleman made exactly the same remark to Mr. Erskine in regard to his countrymen. I noticed a young lady earnestly trying to recollect a painter's name, and she first looked to one corner of the ceiling and then to the opposite corner, arching the one eyebrow on that side; although, of course, there was nothing to be seen there.

In most of the foregoing cases, we can understand how the associated movements were acquired through habit; but with some individuals, certain strange gestures or tricks have arisen in association with certain states of the mind, owing to wholly inexplicable causes, and are undoubtedly inherited.

...

There are other actions which are commonly performed under certain circumstances, independently of habit, and which seem to be due to imitation or some sort of sympathy. Thus persons cutting anything with a pair of scissors may be seen to move their jaws simultaneously with the blades of the scissors. Children learning to write often twist about their tongues as their fingers move, in a ridiculous fashion. When a public singer suddenly becomes a little hoarse, many of those present may be heard, as I have been assured by a gentleman on whom I can rely, to clear their throats; but here habit probably comes into play, as we clear our own throats under similar circumstances. I have also been told that at leaping matches, as the performer makes his spring, many of the spectators, generally men and boys, move their feet; but here again habit probably comes into play, for it is very doubtful whether women would thus act.

...

Associated habitual movements in the lower animals.—I have already given in the case of Man several instances of movements, associated with various states of the mind or body, which are now purposeless, but which were orginally of use, and are still of use under certain circumstances. As this subject is very important for us, I will here give a considerable number of analogous facts, with reference to animals; although many of them are of a very trifling nature. My object is to show that certain movements were originally performed for a definite end, and that, under nearly the same circumstances, they are still pertinaciously performed through habit when not of the least use. That the tendency in most of the following cases is inherited, we may infer from such actions being performed in the same manner by all the individuals, young and old, of the same species. We shall also see that they are excited by the most diversified, often circuitous, and sometimes mistaken associations.

Dogs, when they wish to go to sleep on a carpet or other hard surface, generally turn round and round and scratch the ground with their fore-paws in a senseless manner, as if they intended to trample down the grass and scoop out a hollow, as no doubt their wild parents did, when they lived on open grassy plains or in the woods. Jackals, fennecs, and other allied animals in the Zoological Gardens, treat their straw in this manner; but it is a rather odd circumstance that the keepers, after observing for some months, have never seen the wolves thus behave. A semi-idiotic dog—and an animal in this condition would be particularly liable to follow a senseless habit— was observed by a friend to turn completely round on a carpet thirteen times before going to sleep.

Many carnivorous animals, as they crawl towards their prey and prepare to rush or spring on it, lower their heads and crouch, partly, as it would appear, to hide themselves, and partly to get ready for their rush; and this habit in an exaggerated form has become hereditary in our pointers and setters. Now I have noticed scores of times that when two strange dogs meet on an open road, the one which first sees the other, though at the distance of one or two hundred yards, after the first glance always lowers its head, generally crouches a little, or even lies down; that is, he takes the proper attitude for concealing himself and for making a rush or spring, although the road is quite open and the distance great. Again, dogs of all kinds when intently watching and slowly approaching their prey, frequently keep one of their fore-legs doubled up for a long time, ready for the next cautious step; and this is eminently characteristic of the pointer. But from habit they behave in exactly the same manner whenever their attention is aroused [Fig. 7.1].

I have seen a dog at the foot of a high wall, listening attentively to a sound on the opposite side, with one leg doubled up; and in this case there could have been no intention of making a cautious approach.

Dogs after voiding their excrement often make with all four feet a few scratches backwards, even on a bare stone pavement, as if for the purpose of covering up their excrement with earth, in nearly the same manner as do cats. Wolves and jackals behave in the Zoological Gardens in exactly the same manner, yet, as I am assured by the keepers, neither wolves, jackals, nor foxes, when they have the means of doing so, ever cover up their excrement, any more than do dogs. All these animals, however, bury superfluous food. Hence, if we rightly understand the meaning of the above cat-like habit, of which there can be little doubt, we have a purposeless remnant of an habitual movement, which was originally followed by some remote

Figure 7.1   A dog watching a cat
Small dog watching a cat on a table. From a photograph taken by Mr. Rejlander.

progenitor of the dog-genus for a definite purpose, and which has been retained for a prodigious length of time.

Dogs and jackals take much pleasure in rolling and rubbing their necks and backs on carrion. The odour seems delightful to them, though dogs at least do not eat carrion. Mr. Bartlett has observed wolves for me, and has given them carrion, but has never seen them roll on it. I have heard it remarked, and I believe it to be true, that the larger dogs, which are probably descended from wolves, do not so often roll in carrion as do smaller dogs, which are probably descended from jackals. When a piece of brown biscuit is offered to a terrier of mine and she is not hungry (and I have heard of similar instances), she first tosses it about and worries it, as if it were a rat or other prey; she then repeatedly rolls on it precisely as if it were a piece of carrion, and at last eats it. It would appear that an imaginary relish has to be given to the distasteful morsel; and to effect this the dog acts in his habitual manner, as if the biscuit was a live animal or smelt like carrion, though he knows better than we do that this is not the case. I have seen this same terrier act in the same manner after killing a little bird or mouse.

Dogs scratch themselves by a rapid movement of one of their hind feet; and when their backs are rubbed with a stick, so strong is the habit, that they cannot help rapidly scratching the air or the ground in a useless and ludicrous manner. The terrier just alluded to, when thus scratched with a stick, will sometimes show her delight by another habitual movement, namely, by licking the air as if it were my hand.

Horses scratch themselves by nibbling those parts of their bodies which they can reach with their teeth; but more commonly one horse shows another where he wants to be scratched, and they then nibble each other. A friend whose attention I had called to the subject, observed that when he rubbed his horse's neck, the animal protruded his head, uncovered his teeth, and moved his jaws, exactly as if nibbling another horse's neck, for he could never have nibbled his own neck. If a horse is much tickled, as when curry-combed, his wish to bite something becomes so intolerably strong, that he will clatter his teeth together, and though not vicious, bite his groom. At the same time from habit he closely depresses his ears, so as to protect them from being bitten, as if he were fighting with another horse.

A horse when eager to start on a journey makes the nearest approach which he can to the habitual movement of progression by pawing the ground. Now when horses in their stalls are about to be fed and are eager for their corn, they paw the pavement or the straw. Two of my horses thus behave when they see or hear the corn given to their neighbours. But here we have what may almost be called a true expression, as pawing the ground is universally recognized as a sign of eagerness.

Cats cover up their excrements of both kinds with earth; and my grandfather saw a kitten scraping ashes over a spoonful of pure water spilt on the hearth; so that here an habitual or instinctive action was falsely excited, not by a previous act or by odour, but by eyesight. It is well known that cats dislike wetting their feet, owing, it is probable, to their having aboriginally inhabited the dry country of Egypt; and when they wet their feet they shake them violently. My daughter poured some water into a glass close to the head of a kitten; and it immediately shook its feet in the usual manner; so that here we have an habitual movement falsely excited by an associated sound instead of by the sense of touch.

Kittens, puppies, young pigs and probably many other young animals, alternately push with their fore-feet against the mammary glands of their mothers, to excite a freer secretion of milk, or to make it flow. Now it is very common with young cats, and not at all rare with old cats of the common and Persian breeds (believed by some naturalists to be specifically extinct), when comfortably lying on a warm shawl or other soft substance, to pound it quietly and alternately with their fore-feet; their toes being spread out and claws slightly protruded, precisely as when sucking their mother. That it is the same movement is clearly shown by their often at the same time taking a bit of the shawl into their mouths and sucking it; generally closing their eyes and purring from delight. This curious movement is commonly excited only in association with the sensation of a warm soft surface; but I have seen an old cat, when pleased by having its back scratched, pounding the air with its feet in the same manner; so

180

that this action has almost become the expression of a pleasurable sensation.

. . .

The principle of antithesis

We will now consider our second Principle, that of Antithesis. Certain states of the mind lead, as we have seen in the last chapter, to certain habitual movements which were primarily, or may still be, of service; and we shall find that when a directly opposite state of mind is induced, there is a strong and involuntary tendency to the performance of movements of a directly opposite nature, though these have never been of any service. . . . We are particularly liable to confound conventional or artificial gestures and expressions with those which are innate or universal, and which alone deserve to rank as true expressions, I will in the present chapter almost confine myself to the lower animals.

When a dog approaches a strange dog or man in a savage or hostile frame of mind he walks upright and very stiffly; his head is slightly raised, or not much lowered; the tail is held erect and quite rigid; the hairs bristle, especially along the neck and back; the pricked ears are directed forwards, and the eyes have a fixed stare: [see Figs 7.2 and 7.4]. These actions, as will hereafter be explained, follow from the dog's intention to attack his enemy, and are thus to a large extent intelligible. As he prepares to spring with a savage growl on his enemy, the canine teeth are uncovered, and the ears are pressed close backwards on the head; but with these latter actions, we are not here concerned. Let us now suppose that the dog suddenly discovers that the man whom he is approaching, is not a stranger, but his master [Fig. 7.5]; and let it be observed how completely and instantaneously his whole bearing is reversed. Instead of walking upright, the body sinks downwards or even crouches, and is thrown into flexuous movements; his tail, instead of being held stiff and upright, is lowered and wagged from side to side; his hair instantly becomes smooth; his ears are depressed and drawn backwards, but not closely to the head; and his lips hang loosely. From the drawing back of the ears, the eyelids become elongated, and the eyes no longer appear round and staring. It should be added that the animal is at such times in an excited condition from joy; and nerve-force will be generated in excess, which naturally leads to action of some kind. Not one of the above movements, so clearly expressive of affection, are of the least direct service to the animal. They are explicable, as far as I can see, solely from being in complete opposition or antithesis to the attitude and movements which, from intelligible causes, are assumed when a dog intends to fight, and which consequently are expressive of anger.

Figure 7.2 Hostile dog
Dog approaching another dog with hostile intentions. By Mr. Riviere.

Figure 7.3 Affectionate dog
The same in a humble and affectionate frame of mind. By Mr. Riviere.

Figure 7.4 Hostile dog
Half-bred Shepherd Dog in the same state as in Fig. [7.2]. By Mr. A. May.

Figure 7.5
Affectionate dog
The same caressing his
master. By Mr. A. May.

I request the reader to look at the four accompanying sketches, which have been given in order to recall vividly the appearance of a dog under these two states of mind. It is, however, not a little difficult to represent affection in a dog, whilst caressing his master and wagging his tail, as the essence of the expression lies in the continuous flexuous movements.

We will now turn to the cat. When this animal is threatened by a dog, it arches its back in a surprising manner, erects its hair, opens its mouth and spits. But we are not here concerned with this well-known attitude, expressive of terror combined with anger; we are concerned only with that of rage or anger. This is not often seen, but may be observed when two cats are fighting together; and I have seen it well exhibited by a savage cat whilst plagued by a boy. The attitude is almost exactly the same as that of a tiger disturbed and growling over its food, which every one must have beheld in menageries. The animal assumes a crouching position, with the body extended; and the whole tail, or the tip alone, is lashed or curled from side to side. The hair is not in the least erect. Thus far, the attitude and movements are nearly the same as when the animal is prepared to spring on its prey, and when, no doubt, it feels savage. But when preparing to fight, there is this difference, that the ears are closely pressed backwards; the mouth is partially opened, showing the teeth; the fore feet are occasionally struck out with protruded claws; and the animal occasionally utters a fierce growl. [See Figs 7.6 and 7.7.] All, or almost all, these actions naturally follow (as hereafter to be explained), from the cat's manner and intention of attacking its enemy.

Let us now look at a cat in a directly opposite frame of mind, whilst feeling affectionate and caressing her master; and mark how opposite is her attitude in every respect. She now stands upright with her back slightly arched, which makes the hair appear rather rough, but it does not bristle; her tail, instead of being extended and lashed from side to side, is held quite stiff and perpendicularly upwards; her ears are erect and pointed; her mouth is closed; and she rubs against her master with a purr instead of a growl. Let it further be observed how widely different is the whole bearing of an affectionate cat from that of a dog, when with his body crouching and flexuous, his tail lowered and wagging, and ears depressed, he caresses his master. This contrast in the attitudes and movements of these two carnivorous animals, under the same pleased and affectionate frame of mind, can be explained, as it appears to me, solely by their movements standing in complete antithesis to those which are naturally assumed, when these animals feel savage and are prepared either to fight or to seize their prey.

In these cases of the dog and cat, there is every reason to believe

Figure 7.6   Savage cat
Cat, savage, and prepared to fight, drawn from life by Mr. Wood.

that the gestures both of hostility and affection are innate or inherited; for they are almost identically the same in the different races of the species, and in all the individuals of the same race, both young and old.

I will here give one other instance of antithesis in expression. I formerly possessed a large dog, who, like every other dog, was much pleased to go out walking. He showed his pleasure by trotting gravely before me with high steps, head much raised, moderately erected ears, and tail carried aloft but not stiffly. Not far from my house a path branches off to the right, leading to the hot-house, which I used often to visit for a few moments, to look at my experimental plants. This was always a great disappointment to the dog, as he did not know whether I should continue my walk; and the instantaneous and complete change of expression which came over him, as soon as my body swerved in the least towards the path (and I sometimes tried this as an experiment) was laughable. His look of dejection was known to every member of the family, and was called his *hot-house face*. This consisted in the head drooping much, the whole body sinking a little and remaining motionless; the ears and tail falling suddenly down, but the tail was by no means wagged. With the falling of the ears and of his great chaps, the eyes became much changed in appearance, and I fancied that they looked less bright. His aspect was that of piteous, hopeless dejection; and it was, as I have said, laughable, as the cause was so slight. Every detail in his attitude was in complete opposition to his former joyful yet dignified bearing; and

Figure 7.7   Affectionate cat
Cat in an affectionate frame of mind, by Mr. Wood.

can be explained, as it appears to me, in no other way, except through the principle of antithesis. Had not the change been so instantaneous, I should have attributed it to his lowered spirits affecting, as in the case of man, the nervous system and circulation, and consequently the tone of his whole muscular frame; and this may have been in part the cause.

. . .

The principle of the direct action of the nervous system

We now come to our third Principle, namely, that certain actions, which we recognise as expressive of certain states of the mind, are

186

the direct result of the constitution of the nervous system, and have been from the first independent of the will, and, to a large extent, of habit. When the sensorium is strongly excited nerve-force is generated in excess, and is transmitted in certain directions, dependent on the connection of the nerve-cells, and, as far as the muscular system is concerned, on the nature of the movements which have been habitually practised. Or the supply of nerve-force may, as it appears, be interrupted. Of course every movement which we make is determined by the constitution of the nervous system; but actions performed in obedience to the will, or through habit, or through the principle of antithesis, are here as far as possible excluded. Our present subject is very obscure, but, from its importance, must be discussed at some little length; and it is always advisable to perceive clearly our ignorance.

The most striking case, though a rare and abnormal one, which can be adduced of the direct influence of the nervous system, when strongly affected, on the body, is the loss of colour in the hair, which has occasionally been observed after extreme terror or grief. One authentic instance has been recorded, in the case of a man brought out for execution in India, in which the change of colour was so rapid that it was perceptible to the eye.

Another good case is that of the trembling of the muscles, which is common to man and to many, or most, of the lower animals. Trembling is of no service, often of much disservice, and cannot have been at first acquired through the will, and then rendered habitual in association with any emotion. I am assured by an eminent authority that young children do not tremble, but go into convulsions under the circumstances which would induce excessive trembling in adults. Trembling is excited in different individuals in very different degrees, and by the most diversified causes,—by cold to the surface, before fever-fits, although the temperature of the body is then above the normal standard; in blood-poisoning, delirium tremens, and other diseases; by general failure of power in old age; by exhaustion after excessive fatigue; locally from severe injuries, such as burns; and, in an especial manner, by the passage of a catheter. Of all emotions, fear notoriously is the most apt to induce trembling; but so do occasionally great anger and joy. I remember once seeing a boy who had just shot his first snipe on the wing, and his hands trembled to such a degree from delight, that he could not for some time reload his gun; and I have heard of an exactly similar case with an Australian savage, to whom a gun had been lent. Fine music, from the vague emotions thus excited, causes a shiver to run down the backs of some persons. There seems to be very little in common in the above several physical causes and emotions to account for trembling; and Sir J. Paget, to

187

whom I am indebted for several of the above statements, informs me that the subject is a very obscure one. As trembling is sometimes caused by rage, long before exhaustion can have set in, and as it sometimes accompanies great joy, it would appear that any strong excitement of the nervous system interrupts the steady flow of nerve-force to the muscles.

The manner in which the secretions of the alimentary canal and of certain glands—as the liver, kidneys, or mammæ—are affected by strong emotions, is another excellent instance of the direct action of the sensorium on these organs, independently of the will or of any serviceable associated habit. There is the greatest difference in different persons in the parts which are thus affected, and in the degree of their affection.

The heart, which goes on uninterruptedly beating night and day in so wonderful a manner, is extremely sensitive to external stimulants. The great physiologist, Claude Bernard, has shown how the least excitement of a sensitive nerve reacts on the heart; even when a nerve is touched so slightly that no pain can possibly be felt by the animal under experiment. Hence when the mind is strongly excited, we might expect that it would instantly affect in a direct manner the heart; and this is universally acknowledged and felt to be the case. Claude Bernard also repeatedly insists, and this deserves especial notice, that when the heart is affected it reacts on the brain; and the state of the brain again reacts through the pneumo-gastric nerve on the heart; so that under any excitement there will be much mutual action and reaction between these, the two most important organs of the body.

We now turn from general principle to actual application. The expression of 'decision and determination', 'anger', and 'sneering' illustrate the principle of serviceable associated habit; then 'helplessness' and 'shrugging of shoulders' illustrate the principle of antithesis.

Decision or determination

Decision or determination.—The firm closure of the mouth tends to give an expression of determination or decision to the countenance. No determined man probably ever had an habitually gaping mouth. Hence, also, a small and weak lower jaw, which seems to indicate that the mouth is not habitually and firmly closed, is commonly thought to be characteristic of feebleness of character. A prolonged effort of any kind, whether of body or mind, implies previous determination; and if it can be shown that the mouth is generally closed with firmness

188

before and during a great and continued exertion of the muscular system, then, through the principle of association, the mouth would almost certainly be closed as soon as any determined resolution was taken. Now several observers have noticed that a man, in commencing any violent muscular effort, invariably first distends his lungs with air, and then compresses it by the strong contraction of the muscles of the chest; and to effect this the mouth must be firmly closed. Moreover, as soon as the man is compelled to draw breath, he still keeps his chest as much distended as possible.

Various causes have been assigned for this manner of acting. Sir C. Bell maintains that the chest is distended with air, and is kept distended at such times, in order to give a fixed support to the muscles which are thereto attached. Hence, as he remarks, when two men are engaged in a deadly contest, a terrible silence prevails, broken only by hard stifled breathing. There is silence, because to expel the air in the utterance of any sound would be to relax the support for the muscles of the arms. If an outcry is heard, supposing the struggle to take place in the dark, we at once know that one of the two has given up in despair.

Gratiolet admits that when a man has to struggle with another to his utmost, or has to support a great weight, or to keep for a long time the same forced attitude, it is necessary for him first to make a deep inspiration, and then to cease breathing; but he thinks that Sir C. Bell's explanation is erroneous. He maintains that arrested respiration retards the circulation of the blood, of which I believe there is no doubt, and he adduces some curious evidence from the structure of the lower animals, showing, on the one hand, that a retarded circulation is necessary for prolonged muscular exertion, and, on the other hand, that a rapid circulation is necessary for rapid movements. According to this view, when we commence any great exertion, we close our mouths and stop breathing, in order to retard the circulation of the blood. Gratiolet sums up the subject by saying, "C'est là la vraie théorie de l'effort continu;" but how far this theory is admitted by other physiologists I do not know.

Dr. Piderit accounts for the firm closure of the mouth during strong muscular exertion, on the principle that the influence of the will spreads to other muscles besides those necessarily brought into action in making any particular exertion; and it is natural that the muscles of respiration and of the mouth, from being so habitually used, should be especially liable to be thus acted on. It appears to me that there probably is some truth in this view, for we are apt to press the teeth hard together during violent exertion, and this is not requisite to prevent expiration, whilst the muscles of the chest are strongly contracted.

Lastly, when a man has to perform some delicate and difficult operation, not requiring the exertion of any strength, he nevertheless generally closes his mouth and ceases for a time to breathe; but he acts thus in order that the movements of his chest may not disturb those of his arms. A person, for instance, whilst threading a needle, may be seen to compress his lips and either to stop breathing, or to breathe as quietly as possible. So it was, as formerly stated, with a young and sick chimpanzee, whilst it amused itself by killing flies with its knuckles, as they buzzed about on the window-panes. To perform an action, however trifling, if difficult, implies some amount of previous determination.

There appears nothing improbable in all the above assigned causes having come into play in different degrees, either conjointly or separately, on various occasions. The result would be a well-established habit, now perhaps inherited, of firmly closing the mouth at the commencement and during any violent and prolonged exertion, or any delicate operation. Through the principle of association there would also be a strong tendency towards this same habit, as soon as the mind has resolved on any particular action or line of conduct, even before there was any bodily exertion, or if none was requisite. The habitual and firm closure of the mouth would thus come to show decision of character.

Anger, indignation

Anger, indignation.—These states of the mind differ from rage only in degree, and there is no marked distinction in their characteristic signs. Under moderate anger the action of the heart is a little increased, the colour heightened, and the eyes become bright. The respiration is likewise a little hurried; and as all the muscles serving for this function act in association, the wings of the nostrils are somewhat raised to allow of a free indraught of air; and this is a highly characteristic sign of indignation. The mouth is commonly compressed, and there is almost always a frown on the brow. Instead of the frantic gestures of extreme rage, an indignant man unconsciously throws himself into an attitude ready for attacking or striking his enemy, whom he will perhaps scan from head to foot in defiance. He carries his head erect, with his chest well expanded, and the feet planted firmly on the ground. He holds his arms in various positions, with one or both elbows squared, or with the arms rigidly suspended by his sides. With Europeans the fists are commonly clenched. . . . Any one may see in a mirror, if he will vividly imagine that he has been insulted and demands an explanation in an angry tone of voice, that he suddenly and unconsciously throws himself into some such attitude.

Rage, anger, and indignation are exhibited in nearly the same manner throughout the world; and the following descriptions may be worth giving as evidence of this, and as illustrations of some of the foregoing remarks. There is, however, an exception with respect to clenching the fists, which seems confined chiefly to the men who fight with their fists. With the Australians only one of my informants has seen the fists clenched. All agree about the body being held erect; and all, with two exceptions, state that the brows are heavily contracted. Some of them allude to the firmly-compressed mouth, the distended nostrils and flashing eyes. According to the Rev. Mr. Taplin, rage, with the Australians, is expressed by the lips being protruded, the eyes being widely open; and in the case of the women by their dancing about and casting dust into the air. Another observer speaks of the native men, when enraged, throwing their arms wildly about.

I have received similar accounts, except as to the clenching of the fists, in regard to the Malays of the Malacca peninsula, the Abyssinians, and the natives of South Africa. So it is with the Dakota Indians of North America; and, according to Mr. Matthews, they then hold their heads erect, frown, and often stalk away with long strides. Mr. Bridges states that the Fuegians, when enraged, frequently stamp on the ground, walk distractedly about, sometimes cry and grow pale. The Rev. Mr. Stack watched a New Zealand man and woman quarrelling, and made the following entry in his note-book: "Eyes dilated, body swayed violently backwards and forwards, head inclined forwards, fists clenched, now thrown behind the body, now directed towards each other's faces." Mr. Swinhoe says that my description agrees with what he has seen of the Chinese, excepting that an angry man generally inclines his body towards his antagonist, and pointing at him, pours forth a volley of abuse.

Lastly, with respect to the natives of India, Mr. J. Scott has sent me a full description of their gestures and expression when enraged. Two low-caste Bengalees disputed about a loan. At first they were calm, but soon grew furious and poured forth the grossest abuse on each other's relations and progenitors for many generations past. Their gestures were very different from those of Europeans; for though their chests were expanded and shoulders squared, their arms remained rigidly suspended, with the elbows turned inwards and the hands alternately clenched and opened. Their shoulders were often raised high, and then again lowered. They looked fiercely at each other from under their lowered and strongly wrinkled brows, and their protruded lips were firmly closed. They approached each other, with heads and necks stretched forwards, and pushed, scratched, and grasped at each other. This protrusion of the head and body seems a common gesture with the enraged; and I have noticed it with degraded English women whilst quarrelling violently in the streets. In such cases it may be presumed that neither party expects to receive a blow from the other.

A Bengalee employed in the Botanic Gardens was accused, in the presence of Mr. Scott, by the native overseer of having stolen a valuable plant. He listened silently and scornfully to the accusation; his attitude erect, chest expanded, mouth closed, lips protruding, eyes firmly set and penetrating. He then defiantly maintained his innocence, with upraised and clenched hands, his head being now pushed forwards, with the eyes widely open and eyebrows raised. Mr. Scott also watched two Mechis, in Sikhim, quarrelling about their share of payment. They soon got into a furious passion, and then their bodies became less erect, with their heads pushed forwards; they made grimaces at each other; their shoulders were raised; their arms rigidly bent inwards at the elbows, and their hands spasmodically closed, but not properly clenched. They continually approached and retreated from each other, and often raised their arms as if to strike, but their hands were open, and no blow was given. Mr. Scott made similar observations on the Lepchas whom he often saw quarrelling, and he noticed that they kept their arms rigid and almost parallel to their bodies, with the hands pushed somewhat backwards and partially closed, but not clenched.

Sneering, defiance

Sneering, Defiance: Uncovering the canine tooth on one side. —The expression which I wish here to consider differs but little from that already described, when the lips are retracted and the grinning teeth exposed. The difference consists solely in the upper lip being retracted in such a manner that the canine tooth on one side of the face alone is shown; the face itself being generally a little upturned and half averted from the person causing offence. The other signs of rage are not necessarily present. This expression may occasionally be observed in a person who sneers at or defies another, though there may be no real anger; as when any one is playfully accused of some fault, and answers, "I scorn the imputation." The expression is not a common one, but I have seen it exhibited with perfect distinctness by a lady who was being quizzed by another person. It was described by Parsons as long ago as 1746, with an engraving, showing the uncovered canine on one side. Mr. Rejlander, without my having made any allusion to the subject, asked me whether I had ever noticed this expression, as he had been much struck by it. . . .

The expression of a half-playful sneer graduates into one of great ferocity when, together with a heavily frowning brow and fierce eye, the canine tooth is exposed. A Bengalee boy was accused before Mr.

Scott of some misdeed. The delinquent did not dare to give vent to his wrath in words, but it was plainly shown on his countenance, sometimes by a defiant frown, and sometimes "by a thoroughly canine snarl." When this was exhibited, "the corner of the lip over the eye-tooth, which happened in this case to be large and projecting, was raised on the side of his accuser, a strong frown being still retained on the brow." Sir C. Bell states that the actor Cooke could express the most determined hate "when with the oblique cast of his eyes he drew up the outer part of the upper lip, and discovered a sharp angular tooth."

The uncovering of the canine tooth is the result of a double movement. The angle or corner of the mouth is drawn a little backwards, and at the same time a muscle which runs parallel to and near the nose draws up the outer part of the upper lip, and exposes the canine on this side of the face. The contraction of this muscle makes a distinct furrow on the cheek, and produces strong wrinkles under the eye, especially at its inner corner. The action is the same as that of a snarling dog; and a dog when pretending to fight often draws up the lip on one side alone, namely that facing his antagonist. Our word *sneer* is in fact the same as *snarl*, which was originally *snar*, the *l* "being merely an element implying continuance of action."[1]

I suspect that we see a trace of this same expression in what is called a derisive or sardonic smile. The lips are then kept joined or almost joined, but one corner of the mouth is retracted on the side towards the derided person; and this drawing back of the corner is part of a true sneer. Although some persons smile more on one side of their face than on the other, it is not easy to understand why in cases of derision the smile, if a real one, should so commonly be confined to one side. I have also on these occasions noticed a slight twitching of the muscle which draws up the outer part of the upper lip; and this movement, if fully carried out, would have uncovered the canine, and would have produced a true sneer.

Mr. Bulmer, an Australian missionary in a remote part of Gipps' Land, says, in answer to my query about the uncovering of the canine on one side, "I find that the natives in snarling at each other speak with the teeth closed, the upper lip drawn to one side, and a general angry expression of face; but they look direct at the person addressed." Three other observers in Australia, one in Abyssinia, and one in China, answer my query on this head in the affirmative; but as the expression is rare, and as they enter into no details, I am afraid of implicitly trusting them. It is, however, by no means improbable that this animal-like expression may be more common with savages than with civilized races. Mr. Geach is an observer who may be fully trusted, and he has observed it on one occasion in a Malay in the interior of Malacca. The Rev. S. O. Glenie answers, "We have observed this expression with the natives of Ceylon, but not often." Lastly, in North America, Dr. Rothrock had seen it with some wild Indians, and often in a tribe adjoining the Atnahs.

Although the upper lip is certainly sometimes raised on one side alone in sneering at or defying any one, I do not know that this is always the case, for the face is commonly half averted, and the expression is often momentary. The movement being confined to one side may not be an essential part of the expression, but may depend on the proper muscles being incapable of movement excepting on one side. I asked four persons to endeavour to act voluntarily in this manner; two could expose the canine only on the left side, one only on the right side, and the fourth on neither side. Nevertheless it is by no means certain that these same persons, if defying any one in earnest, would not unconsciously have uncovered their canine tooth on the side, whichever it might be, towards the offender. For we have seen that some persons cannot voluntarily make their eyebrows oblique, yet instantly act in this manner when affected by any real, although most trifling, cause of distress. The power of voluntarily uncovering the canine on one side of the face being thus often wholly lost, indicates that it is a rarely used and almost abortive action. It is indeed a surprising fact that man should possess the power, or should exhibit any tendency to its use; for Mr. Sutton has never noticed a snarling action in our nearest allies, namely, the monkeys in the Zoological Gardens, and he is positive that the baboons, though furnished with great canines, never act thus, but uncover all their teeth when feeling savage and ready for an attack. Whether the adult anthropomorphous apes, in the males of whom the canines are much larger than in the females, uncover them when prepared to fight, is not known.

The expression here considered, whether that of a playful sneer or ferocious snarl, is one of the most curious which occurs in man. It reveals his animal descent; for no one, even if rolling on the ground in a deadly grapple with an enemy, and attempting to bite him, would try to use his canine teeth more than his other teeth. We may readily believe from our affinity to the anthropomorphous apes that our male semi-human progenitors possessed great canine teeth, and men are now occasionally born having them of unusually large size, with interspaces in the opposite jaw for their reception. We may further suspect, notwithstanding that we have no support from analogy, that our semi-human progenitors uncovered their canine teeth when prepared for battle, as we still do when feeling ferocious, or when merely sneering at or defying some one, without any intention of making a real attack with our teeth.

Shrugging the shoulders

Helplessness, Impotence: Shrugging the shoulders. —When a man wishes to show that he cannot do something, or prevent something being done, he often raises with a quick movement both shoulders. At the same time, if the whole gesture is completed, he bends his elbows closely inwards, raises his open hands, turning them outwards, with the fingers separated. The head is often thrown a little on one side; the eyebrows are elevated, and this causes wrinkles across the forehead. The mouth is generally opened. I may mention, in order to show how unconsciously the features are thus acted on, that though I had often intentionally shrugged my shoulders to observe how my arms were placed, I was not at all aware that my eyebrows were raised and mouth opened, until I looked at myself in a glass; and since then I have noticed the same movements in the faces of others. . . .

Englishmen are much less demonstrative than the men of most other European nations, and they shrug their shoulders far less frequently and energetically than Frenchmen or Italians do. The gesture varies in all degrees from the complex movement, just described, to only a momentary and scarcely perceptible raising of both shoulders; or, as I have noticed in a lady sitting in an arm-chair, to the mere turning slightly outwards of the open hands with separated fingers. I have never seen very young English children shrug their shoulders, but the following case was observed with care by a medical professor and excellent observer, and has been communicated to me by him. The father of this gentleman was a Parisian, and his mother a Scotch lady. His wife is of British extraction on both sides, and my informant does not believe that she ever shrugged her shoulders in her life. His children have been reared in England, and the nursemaid is a thorough English-woman, who has never been seen to shrug her shoulders. Now, his eldest daughter was observed to shrug her shoulders at the age of between sixteen and eighteen months; her mother exclaiming at the time, "Look at the little French girl shrugging her shoulders!" At first she often acted thus, sometimes throwing her head a little backwards and on one side, but she did not, as far as was observed, move her elbows and hands in the usual manner. The habit gradually wore away, and now, when she is a little over four years old, she is never seen to act thus. The father is told that he sometimes shrugs his shoulders, especially when arguing with any one; but it is extremely improbable that his daughter should have imitated him at so early an age; for, as he remarks, she could not possibly have often seen this gesture in him. Moreover, if the habit had been acquired through imitation, it is not probable that it would so soon have been spontaneously discontinued by this child, and, as we shall immediately see, by a second child, though the father still lived with his family. This little girl, it may be added, resembles her Parisian grandfather in countenance to an almost absurd degree. She also presents another and very curious resemblance to him, namely, by practising a singular trick. When she impatiently wants something, she holds out her little hand, and rapidly rubs the thumb against the index and middle finger: now this same trick was frequently performed under the same circumstances by her grandfather.

This gentleman's second daughter also shrugged her shoulders before the age of eighteen months, and afterwards discontinued the habit. It is of course possible that she may have imitated her elder sister; but she continued it after her sister had lost the habit. She at first resembled her Parisian grandfather in a less degree than did her sister at the same age, but now in a greater degree. She likewise practises to the present time the peculiar habit of rubbing together, when impatient, her thumb and two of her fore-fingers.

In this latter case we have a good instance . . . of the inheritance of a trick or gesture; for no one, I presume, will attribute to mere coincidence so peculiar a habit as this, which was common to the grandfather and his two grandchildren who had never seen him.

Considering all the circumstances with reference to these children shrugging their shoulders, it can hardly be doubted that they have inherited the habit from their French progenitors, although they have only one quarter French blood in their veins, and although their grandfather did not often shrug his shoulders. There is nothing very unusual, though the fact is interesting, in these children having gained by inheritance a habit during early youth, and then discontinuing it; for it is of frequent occurrence with many kinds of animals that certain characters are retained for a period by the young, and are then lost.

As it appeared to me at one time improbable in a high degree that so complex a gesture as shrugging the shoulders, together with the accompanying movements, should be innate, I was anxious to ascertain whether the blind and deaf Laura Bridgman, who could not have learnt the habit by imitation, practised it. And I have heard, through Dr. Innes, from a lady who has lately had charge of her, that she does shrug her shoulders, turn in her elbows, and raise her eyebrows in the same manner as other people, and under the same circumstances. I was also anxious to learn whether this gesture was practised by the various races of man, especially by those who never have had much intercourse with Europeans. We shall see that they act in this manner; but it appears that the gesture is sometimes confined to merely raising or shrugging the shoulders, without the other movements.

Mr. Scott has frequently seen this gesture in the Bengalees and Dhangars (the latter constituting a distinct race) who are employed in the Botanic Garden at Calcutta; when, for instance, they have declared that they could not do some work, such as lifting a heavy weight. He ordered a Bengalee to climb a lofty tree; but the man, with a shrug of his shoulders and a lateral shake of his head, said he could not. Mr. Scott knowing that the man was lazy, thought he could, and insisted on his trying. His face now became pale, his arms dropped to his sides, his mouth and eyes were widely opened, and again surveying the tree, he looked askant at Mr. Scott, shrugged his shoulders, inverted his elbows, extended his open hands, and with a few quick lateral shakes of the head declared his inability. Mr. H. Erskine has likewise seen the natives of India shrugging their shoulders; but he has never seen the elbows turned so much inwards as with us; and whilst shrugging their shoulders they sometimes lay their uncrossed hands on their breasts.

. . .

These statements,[2] relating to Europeans, Hindoos, the hill-tribes of India, Malays, Micronesians, Abyssinians, Arabs, Negroes, Indians of North America, and apparently to the Australians—many of these natives having had scarcely any intercourse with Europeans—are sufficient to show that shrugging the shoulders, accompanied in some cases by the other proper movements, is a gesture natural to mankind.

This gesture implies an unintentional or unavoidable action on our own part, or one that we cannot perform; or an action performed by another person which we cannot prevent. It accompanies such speeches as, "It was not my fault;" "It is impossible for me to grant this favour;" "He must follow his own course, I cannot stop him." Shrugging the shoulders likewise expresses patience, or the absence of any intention to resist. Hence the muscles which raise the shoulders are sometimes called, as I have been informed by an artist, "the patience muscles." Shylock the Jew, says,

> "Signor Antonio, many a time and oft
>   In the Rialto have you rated me
>   About my monies and usances;
>   Still have I borne it with a patient shrug."
> *Merchant of Venice*, act i. sc.3.

Sir C. Bell has given a life-like figure of a man, who is shrinking back from some terrible danger, and is on the point of screaming out in abject terror. He is represented with his shoulders lifted up almost to his ears; and this at once declares that there is no thought of resistance.

As shrugging the shoulders generally implies "I cannot do this or that," so by a slight change, it sometimes implies "I won't do it." The movement then expresses a dogged determination not to act. Olmsted describes an Indian in Texas as giving a great shrug to his shoulders, when he was informed that a party of men were Germans and not Americans, thus expressing that he would have nothing to do with them. Sulky and obstinate children may be seen with both their shoulders raised high up; but this movement is not associated with the others which generally accompany a true shrug. An excellent observer in describing a young man who was determined not to yield to his father's desire, says, "He thrust his hands deep down into his pockets, and set up his shoulders to his ears, which was a good warning that, come right or wrong, this rock should fly from its firm base as soon as Jack would; and that any remonstrance on the subject was purely futile." As soon as the son got his own way, he "put his shoulders into their natural position."

Resignation is sometimes shown by the open hands being placed, one over the other, on the lower part of the body. I should not have thought this little gesture worth even a passing notice, had not Dr. W. Ogle remarked to me that he had two or three times observed it in patients who were preparing for operations under chloroform. They exhibited no great fear, but seemed to declare by this posture of their hands, that they had made up their minds, and were resigned to the inevitable.

We may now inquire why men in all parts of the world when they feel,—whether or not they wish to show this feeling,—that they cannot or will not do something, or will not resist something if done by another, shrug their shoulders, at the same time often bending in their elbows, showing the palms of their hands with extended fingers, often throwing their heads a little on one side, raising their eyebrows, and opening their mouths. These states of the mind are either simply passive, or show a determination not to act. None of the above movements are of the least service. The explanation lies, I cannot doubt, in the principle of unconscious antithesis. This principle here seems to come into play as clearly as in the case of a dog, who, when feeling savage, puts himself in the proper attitude for attacking and for making himself appear terrible to his enemy; but as soon as he feels affectionate, throws his whole body into a directly opposite attitude, though this is of no direct use to him.

Let it be observed how an indignant man, who resents, and will not submit to some injury, holds his head erect, squares his shoulders, and expands his chest. He often clenches his fists, and puts one or both arms in the proper position for attack or defence, with the muscles of his limbs rigid. He frowns,—that is, he contracts and lowers his brows,—and, being determined, closes his mouth. The actions and attitude of a helpless man are, in every one of these respects, exactly the reverse. . . .

In accordance with the fact that squaring the elbows and clenching the fists are gestures by no means universal with the men of all races, when they feel indignant and are prepared to attack their enemy, so it appears that a helpless or apologetic frame of mind is expressed in many parts of the world by merely shrugging the shoulders, without turning inwards the elbows and opening the hands. The man or child who is obstinate, or one who is resigned to some great misfortune, has in neither case any idea of resistance by active means; and he expresses this state of mind, by simply keeping his shoulders raised; or he may possibly fold his arms across his breast.

. . .

Notes

1   Darwin here quotes from Hensleigh Wedgwood's *Dictionary of English etymology* (1865). Hensleigh Wedgwood was Darwin's cousin and brother-in-law. His etymological writings have found little favour with later philologists; according to a biographer, he 'had an extraordinary command of linguistic materials and great natural sagacity, marred by an imperfect acquaintance with the discoveries of philological science', which is about as rude a remark as can be made about a philologist.

2   I have omitted Darwin's evidence from the Malays to the Indians of North America; it makes the same point as the included Indian instance.

---------- *Chapter eight* ----------

The different forms of
flowers on plants of
the same species
(1877)

The *Forms of flowers* is the third of Darwin's three books on cross-fertilisation in flowers. He had first studied the fertilisation of flowers by insects in the summer of 1838, and 'attended to the subject during more or less every subsequent summer.' He worked at the subject off and on for almost 40 years. He made discoveries of two main kinds. By experimentally self and cross-fertilising different flowers of a species, he found that self-fertilised flowers were much less productive than cross-fertilised ones. He summarised his results in a whole book, the *Effects of cross and self fertilisation in the vegetable kingdom* (1876). Cross-fertilisation, then, was advantageous, and natural selection should favour adaptations that cause it to take place.

Darwin not only showed that it was advantageous; he also showed how certain flowers are in practice adapted to prevent self-fertilisation and to encourage cross-fertilisation. Such was the theme of his first botanical book, *On the various contrivances by which British and foreign orchids are fertilised by insects*, which in 1862 was his first publication after the *Origin*. The *Effects of cross and self fertilisation*, he explained, 'will form a complement to the Fertilisation of Orchids, in which I showed how perfect were the means for cross-fertilisation, and here I shall show how important are the results.'

As it turned out, the subject of his work on heterostyly (as it is now called) was the same. Heterostyly means that different flowers of the same species contain styles of different lengths. (The style is the tubular part of the female reproductive system of a flower that connects the stigma, where the pollen is received, and the ovule, where the seed is fertilised.) Primroses (*Primula* sp.), for instance, have styles of two distinct lengths; the purple loosestrife (*Lythrum salicaria*) has three. Darwin studied fertilisation in several of these heterostylous species. The first was the cowslip *Primula veris*, and I have extracted that work below. It has a long and a short-styled flower type (Fig. 8.1).

Initially Darwin thought the two types were two separate sexes. The long-styled plants, he thought, were female, and produced seed, and the short-styled ones male, and produced only pollen. He was soon forced to abandon that hypothesis. Both types, he found, produce seeds. As we shall see, he went on to do experiments that revealed the true function of the heterostylous condition. Heterostyly acts to ensure cross-fertilisation, because a short-styled flower has to be fertilised by a long-styled one; it cannot be fertilised by another short-styled flower, such as itself. The solution delighted Darwin. 'No little discovery of mine ever gave me so much pleasure as the making out the meaning of heterostyled flowers.' Let us now turn to that most pleasurable discovery.

Heterostyled dimorphic plants: primulaceæ

It has long been known to botanists that the common Cowslip (*Primula veris*, Brit. Flora, var. *officinalis*, Lin.) exists under two forms, about equally numerous, which obviously differ from each other in the length of their pistils and stamens. This difference has hitherto been looked at as a case of mere variability, but this view, as we shall presently see, is far from the true one. Florists who cultivate the Polyanthus and Auricula have long been aware of the two kinds of flowers, and they call the plants which display the globular stigma at the mouth of the corolla, "pin-headed" or "pin-eyed," and those which display the anthers, "thrum-eyed." I will designate the two forms as the long-styled and short-styled.

The pistil in the long-styled form is almost exactly twice as long as that of the short-styled. The stigma stands in the mouth of the corolla or projects just above it, and is thus externally visible. It stands high above the anthers, which are situated halfway down the tube and cannot be easily seen. In the short-styled form the anthers are attached near the mouth of the tube, and therefore stand above the stigma, which is seated in about the middle of the tubular corolla. The corolla itself is of a different shape in the two forms; the throat or expanded portion above the attachment of the anthers being much longer in the long-styled than in the short-styled form. Village children notice this difference, as they can best make necklaces by threading and slipping the corollas of the long-styled flowers into one another. But there are much more important differences. The stigma in the long-styled form is globular; in the short-styled it is depressed on the summit, so that the longitudinal axis of the former is sometimes nearly double that of the latter. Although somewhat variable in shape, one difference is persistent, namely, in roughness: in some specimens carefully compared, the papillæ which render the stigma rough were in the long-styled form from twice to thrice as long as in the short-styled. The anthers do not differ in size in the two forms, which I mention because this is the case with some hetero-styled plants. The most remarkable difference is in the pollen-grains. I measured with the micrometer many specimens, both dry and wet, taken from plants growing in different situations, and always found a

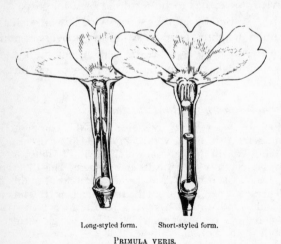

Long-styled form. Short-styled form.

PRIMULA VERIS.

Figure 8.1 Heterostyly of *Primula veris*

palpable difference. The grains distended with water from the short-styled flowers were about ·038 mm. ($^{10-11}$/$_{7000}$ of an inch) in diameter, whilst those from the long-styled were about ·0254 mm. (^{7}/$_{7000}$ of an inch), which is in the ratio of 100 to 67. The pollen-grains, therefore from the longer stamens of the short-styled form are plainly larger than those from the shorter stamens of the long-styled. When examined dry, the smaller grains are seen under a low power to be more transparent than the larger grains, and apparently in a greater degree than can be accounted for by their less diameter. There is also a difference in shape, the grains from the short-styled plants being nearly spherical, those from the long-styled being oblong with the angles rounded; this difference disappears when the grains are distended with water. The long-styled plants generally tend to flower a little before the short-styled: for instance I had twelve plants of each form growing in separate pots and treated in every respect alike; and at the time when only a single short-styled plant was in flower, seven of the long-styled had expanded their flowers.

We shall, also, presently see that the short-styled plants produce more seed than the long-styled. It is remarkable, according to Prof. Oliver, that the ovules in the unexpanded and unimpregnated flowers of the latter are considerably larger than those of the short-styled flowers; and this I suppose is connected with the long-styled flowers producing fewer seeds, so that the ovules have more space and nourishment for rapid development.

To sum up the differences:—The long-styled plants have a much longer pistil, with a globular and much rougher stigma, standing high above the anthers. The stamens are short; the grains of pollen smaller and oblong in shape. The upper half of the tube of the corolla is more expanded. The number of seeds produced is smaller and the ovules larger. The plants tend to flower first.

The short-styled plants have a short pistil, half the length of the tube of the corolla, with a smooth depressed stigma standing beneath the anthers. The stamens are long; the grains of pollen are spherical and larger. The tube of the corolla is of uniform diameter except close to the upper end. The number of seeds produced is larger.

I have examined a large number of flowers; and though the shape of the stigma and the length of the pistil both vary, especially in the short-styled form, I have never met with any transitional states between the two forms in plants growing in a state of nature.[1] There is never the slightest doubt under which form a plant ought to be classed. The two kinds of flowers are never found on the same individual plant. I marked many Cowslips and Primroses, and on the following year all retained the same character, as did some in my garden which flowered out of their proper season in the autumn. Mr.

W. Wooler, of Darlington, however, informs us that he has seen early blossoms on the Polyanthus,[2] which were not long-styled, but became so later in the season. Possibly in this case the pistils may not have been fully developed during the early spring. An excellent proof of the permanence of the two forms may be seen in nursery-gardens, where choice varieties of the Polyanthus are propagated by division; and I found whole beds of several varieties, each consisting exclusively of the one or the other form. The two forms exist in the wild state in about equal numbers: I collected 522 umbels from plants growing in several stations, taking a single umbel from each plant; and 241 were long-styled, and 281 short-styled. No difference in tint or size could be perceived in the two great masses of flowers.

We shall presently see that most of the species of Primula exist under two analogous forms; and it may be asked what is the meaning of the above-described important differences in their structure? The question seems well worthy of careful investigation, and I will give my observations on the cowslip in detail. The first idea which naturally occurred to me was, that this species was tending towards a diœcious condition; that the long-styled plants, with their longer pistils, rougher stigmas, and smaller pollen-grains, were more feminine in nature, and would produce more seed;—that the short-styled plants, with their shorter pistils, longer stamens and larger pollen-grains, were more masculine in nature. Accordingly, in 1860, I marked a few cowslips of both forms growing in my garden, and others growing in an open field, and others in a shady wood, and gathered and weighed the seed. In all the lots the short-styled plants yielded, contrary to my expectation, most seed. Taking the lots together, the following is the result[3]:—

TABLE 1

	Number of Plants	Number of Umbels produced	Number of Capsules produced	Weight of Seed in grains
Short-styled cowslips	9	33	199	83
Long-styled cowslips	13	51	261	91

If we compare the weight from an equal number of plants, and from an equal number of umbels, and from an equal number of capsules of the two forms, we get the following results[4]:—

Primulaceæ

TABLE 2

	Number of Plants	Weight of Seed in grains	Number of Umbels	Weight of Seed	Number of Capsules	Weight of Seed in grains
Short-styled cowslips	10	92	100	251	100	41
Long-styled cowslips	10	70	100	178	100	34

So that, by all these standards of comparison, the short-styled form is the more fertile; if we take the number of umbels (which is the fairest standard, for large and small plants are thus equalised), the short-styled plants produce more seed than the long-styled, in the proportion of nearly four to three.

. . .

We may safely conclude that the short-styled form is more productive than the long-styled form, and the same result holds good with some other species of Primula. Consequently my anticipation that the plants with longer pistils, rougher stigmas, shorter stamens and smaller pollen-grains, would prove to be more feminine in nature, is exactly the reverse of the truth.[5]

In 1860 a few umbels on some plants of both the long-styled and short-styled form, which had been covered by a net, did not produce any seed, though other umbels on the same plants, artificially fertilised, produced an abundance of seed; and this fact shows that the mere covering in itself was not injurious. Accordingly, in 1861, several plants were similarly covered just before they expanded their flowers; these turned out as follows:—

TABLE 5

	Number of Plants	Number of Umbels produced	Product of Seed
Short-styled	6	24	1·3 grain weight of seed, or about 50 in number
Long-styled	18	74	Not one seed

Judging from the exposed plants which grew all round in the same bed, and had been treated in the same manner, excepting that they had been exposed to the visits of insects, the above six short-styled plants ought to have produced 92 grains' weight of seed instead of only 1·3; and the eighteen long-styled plants, which produced not one seed, ought to have produced above 200 grains' weight. The production of a few seeds by the short-styled plants was probably due to the action of Thrips or of some other minute insect. It is scarcely necessary to give any additional evidence, but I may add that ten pots of Polyanthuses and Cowslips of both forms, protected from insects in my greenhouse, did not set one pod, though artificially fertilised flowers in other pots produced an abundance. We thus see that the visits of insects are absolutely necessary for the fertilisation of *Primula veris*. If the corolla of the long-styled form had dropped off, instead of remaining attached in a withered state to the ovarium, the anthers attached to the lower part of the tube with some pollen still adhering to them would have been dragged over the stigma, and the flowers would have been partially self-fertilised, as is the case with *Primula Sinensis* through this means. It is a rather curious fact that so trifling a difference as the falling-off of the withered corolla, should make a very great difference in the number of seeds produced by a plant if its flowers are not visited by insects.

The flowers of the Cowslip and of the other species of the genus secrete plenty of nectar; and I have often seen humble-bees, especially *B. hortorum* and *muscorum*, sucking the former in a proper manner, though they sometimes bite holes through the corolla. No doubt moths likewise visit the flowers, as one of my sons caught *Cucullia verbasci* in the act. The pollen readily adheres to any thin object which is inserted into a flower. The anthers in the one form stand nearly, but not exactly, on a level with the stigma of the other; for the distance between the anthers and stigma in the short-styled form is greater than that in the long-styled, in the ratio of 100 to 90. This difference is the result of the anthers in the long-styled form standing rather higher in the tube than does the stigma in the short-styled, and this favours their pollen being deposited on it. It follows from the position of the organs that if the proboscis of a dead humble-bee, or a thick bristle or rough needle, be pushed down the corolla, first of one form and then of the other, as an insect would do in visiting the two forms growing mingled together, pollen from the long-stamened form adheres round the base of the object, and is left with certainty on the stigma of the long-styled form; whilst pollen from the short stamens of the long-styled form adheres a little way above the extremity of the object, and some is generally left on the stigma of the other form. In accordance with this observation I found that the two kinds of pollen, which could easily be recognised under the microscope, adhered in this manner to the proboscides of the two species of humble-bees and of the moth, which were caught visiting the flowers; but some small grains were mingled with the larger grains round the base of the proboscis, and conversely some large grains with the small grains near the extremity of the proboscis. Thus pollen will be regularly carried from the one form to the other, and they will reciprocally fertilise one another. Nevertheless an insect in withdrawing its proboscis from the corolla of the long-styled form cannot fail occasionally to leave pollen from the same flower on the stigma; and in this case there might be self-fertilisation. But this will be much more likely to occur with the short-styled form; for when I inserted a bristle or other such object into the corolla of this form, and had, therefore, to pass it down between the anthers seated round the mouth of the corolla, some pollen was almost invariably carried down and left on the stigma. Minute insects, such as Thrips, which sometimes haunt the flowers, would likewise be apt to cause the self-fertilisation of both forms.

The several foregoing facts led me to try the effects of the two kinds of pollen on the stigmas of the two forms. Four essentially different unions are possible; namely, the fertilisation of the stigma of the long-styled form by its own-form pollen, and by that of the short-styled; and the stigma of the short-styled form by its own-form pollen, and by that of the long-styled. The fertilisation of either form with pollen from the other form may be conveniently called a *legitimate union*, from reasons hereafter to be made clear; and that of either form with its own-form pollen an *illegitimate union*. . . . The illegitimate unions of both forms might have been tried in three ways; for a flower of either form may be fertilised with pollen from the same flower, or with that from another flower on the same plant, or with that from a distinct plant of the same form. But to make my experiments perfectly fair, and to avoid any evil result from self-fertilisation or too close interbreeding, I have invariably employed pollen from a distinct plant of the same form for the illegitimate unions of all the species; and therefore it may be observed that I have used the term "own-form pollen" in speaking of such unions. The several plants in all my experiments were treated in exactly the same manner, and were carefully protected by fine nets from the access of insects, excepting Thrips, which it is impossible to exclude. I performed all the manipulations myself, and weighed the seeds in a chemical balance; but during many subsequent trials I followed the more accurate plan of counting the seeds. Some of the capsules contained no seeds, or only two or three, and these are excluded in the column headed "good capsules" in several of the following tables:—

TABLE 6

Primula veris

Nature of the Union	Number of Flowers fertilised	Total Number of Capsules produced	Number of good Capsules	Weight of Seed in grains	Calculated Weight of Seed from 100 good Capsules
Long-styled by pollen of short-styled. Legitimate union	22	15	14	8·8	62
Long-styled by own-form pollen. Illegitimate union	20	8	5	2·1	42
Short-styled by pollen of long-styled. Legitimate union	13	12	11	4·9	44
Short-styled by own-form pollen. Illegitimate union	15	8	6	1·8	30
Summary The two legitimate unions	35	27	25	13·7	54
The two illegitimate unions	35	16	11	3.9	35

The results may be given in another form (Table 7) by comparing, first, the number of capsules, whether good or bad, or of the good alone, produced by 100 flowers of both forms when legitimately and illegitimately fertilised; secondly, by comparing the weight of seed in 100 of these capsules, whether good or bad; or, thirdly, in 100 of the good capsules.

TABLE 7

Nature of the Union	Number of Flowers fertilised	Number of Capsules	Number of good Capsules	Weight of Seed in grains	Number of Capsules	Weight of Seed in grains	Number of good Capsules	Weight of Seed in grains
The two legitimate unions	100	77	71	39	100	50	100	54
The two illegitimate unions	100	45	31	11	100	24	100	35

We here see that the long-styled flowers fertilised with pollen from the short-styled yield more capsules, especially good ones (i.e. containing more than one or two seeds), and that these capsules contain a greater proportional weight of seeds than do the flowers of the long-styled when fertilised with pollen from a distinct plant of the same form. So it is with the short-styled flowers, if treated in an analogous manner. Therefore I have called the former method of fertilisation a legitimate union, and the latter, as it fails to yield the full complement of capsules and seeds, an illegitimate union. These two kinds of union are graphically represented in [Fig. 8·2].

If we consider the results of the two legitimate unions taken together and the two illegitimate ones, as shown in Table 7, we see that the former compared with the latter yielded capsules, whether containing many seeds or only a few, in the proportion of 77 to 45, or as 100 to 58. But the inferiority of the illegitimate unions is here perhaps too great, for on a subsequent occasion 100 long-styled and short-styled flowers were illegitimately fertilised, and they together yielded 53 capsules: therefore the rate of 77 to 53, or as 100 to 69, is a fairer one than that of 100 to 58. Returning to Table 7, if we consider only the good capsules, those from the two legitimate unions were to those from the two illegitimate in number as 71 to 31, or as 100 to 44. Again, if we take an equal number of capsules, whether good or bad, from the legitimately and illegitimately fertilised flowers, we find that

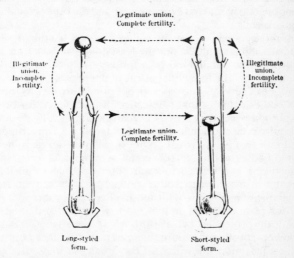

Figure 8.2   **Legitimate and illegitimate fertilisation in *Primula***

the former contained seeds by weight compared with the latter as 50 to 24, or as 100 to 48; but if all the poor capsules are rejected, of which many were produced by the illegitimately fertilised flowers, the proportion is 54 to 35, or as 100 to 65. In this and all other cases, the relative fertility of the two kinds of union can, I think, be judged of more truly by the average number of seeds per capsule than by the proportion of flowers which yield capsules. The two methods might have been combined by giving the average number of seeds produced by all the flowers which were fertilised, whether they yielded capsules or not; but I have thought that it would be more instructive always to show separately the proportion of flowers which produced capsules, and the average number of apparently good seeds which the capsules contained.

Flowers legitimately fertilised set seeds under conditions which cause the almost complete failure of illegitimately fertilised flowers. Thus in the spring of 1862 forty flowers were fertilised at the same time in both ways. The plants were accidentally exposed in the greenhouse to too hot a sun, and a large number of umbels perished. Some, however, remained in moderately good health, and on these there were twelve flowers which had been fertilised legitimately, and eleven which had been fertilised illegitimately. The twelve legitimate unions yielded seven fine capsules, containing on an average each 57·3 good seeds; whilst the eleven illegitimate unions yielded only two capsules, of which one contained 39 seeds, but so poor, that I do not suppose one would have germinated, and the other contained 17 fairly good seeds.

From the facts now given the superiority of a legitimate over an illegitimate union admits of not the least doubt; and we have here a case to which no parallel exists in the vegetable or, indeed, in the animal kingdom. The individual plants of the present species, and as we shall see of several other species of Primula, are divided into two sets or bodies, which cannot be called distinct sexes, for both are hermaphrodites; yet they are to a certain extent sexually distinct, for they require reciprocal union for perfect fertility. As quadrupeds are divided into two nearly equal bodies of different sexes, so here we have two bodies, approximately equal in number, differing in their sexual powers and related to each other like males and females. There are many hermaphrodite animals which cannot fertilise themselves, but must unite with another hermaphrodite. So it is with numerous plants; for the pollen is often mature and shed, or is mechanically protruded, before the flower's own stigma is ready; and such flowers absolutely require the presence of another hermaphrodite for sexual union. But with the Cowslip; and various other species of Primula there is this wide difference, that one individual, though it can fertilise itself imperfectly, must unite with another individual for full fertility; it cannot, however, unite with any other individual in the same manner as an hermaphrodite plant can unite with any other one of the same species; or as one snail or earth-worm can unite with any other hermaphrodite individual. On the contrary, an individual belonging to one form of the Cowslip in order to be perfectly fertile must unite with one of the other form, just as a male quadruped must and can unite only with the female.

I have spoken of the legitimate unions as being fully fertile; and I am fully justified in doing so, for flowers artificially fertilised in this manner yielded rather more seeds than plants naturally fertilised in a state of nature. The excess may be attributed to the plants having been grown separately in good soil. With respect to the illegitimate unions, we shall best appreciate their degree of lessened fertility by the following facts. Gärtner estimated the sterility of the unions between distinct species, in a manner which allows of a strict comparison with the results of the legitimate and illegitimate unions of Primula. With *P. veris*, for every 100 seeds yielded by the two legitimate unions, only 64 were yielded by an equal number of good capsules from the two illegitimate unions. With *P. Sinensis*, as we shall hereafter see, the proportion was nearly the same—namely, as 100 to 62. Now Gärtner has shown that, on the calculation of *Verbascum lychnitis* yielding with its own pollen 100 seeds, it yielded when fertilised by the pollen of *V. Phœniceum* 90 seeds; by the pollen of *V. nigrum*, 63 seeds; by that of *V. blattaria*, 62 seeds. So again, *Dianthus barbatus* fertilised by the pollen of *D. superbus* yielded 81 seeds, and by the pollen of *D. Japonicus* 66 seeds, relatively to the 100 seeds produced by its own pollen. We thus see—and the fact is highly remarkable—that with Primula the illegitimate unions relatively to the legitimate are more sterile than crosses between distinct species of other genera relatively to their pure unions. Mr. Scott has given a still more striking illustration of the same fact: he crossed *Primula auricula* with pollen of four other species (*P. Palinuri, viscosa, hirsuta,* and *verticillata*), and these hybrid unions yielded a larger average number of seeds than did *P. auricula* when fertilised illegitimately with its own-form pollen.

The benefit which heterostyled dimorphic plants derive from the existence of the two forms is sufficiently obvious, namely, the intercrossing of distinct plants being thus ensured.[6] Nothing can be better adapted for this end than the relative positions of the anthers and stigmas in the two forms, as shown in [Fig. 8.2]. . . . No doubt pollen will occasionally be placed by insects or fall on the stigma of the same flower; and if cross-fertilisation fails, such self-fertilisation will be advantageous to the plant, as it will thus be saved from complete barrenness. But the advantage is not so great as might at first be thought, for the seedlings from illegitimate unions do not generally consist of both forms, but all belong to the parent form; they are, moreover, in some degree weakly in constitution, as will be shown in a future chapter. If, however, a flower's own pollen should first be placed by insects or fall on the stigma, it by no means follows that cross-fertilisation will be thus prevented. It is well known that if pollen from a distinct species be placed on the stigma of a plant, and some hours afterwards its own pollen be placed on it, the latter will be prepotent and will quite obliterate any effect from the foreign pollen; and there can hardly be a doubt that with heterostyled dimorphic plants, pollen from the other form will obliterate the effects of pollen from the same form, even when this has been placed on the stigma a considerable time before. To test this belief, I placed on several stigmas of a long-styled cowslip plenty of pollen from the same plant, and after twenty-four hours added some from a short-styled dark-red Polyanthus, which is a variety of the Cowslip. From the flowers thus treated 30 seedlings were raised, and all these, without exception, bore reddish flowers; so that the effect of pollen from the same form, though placed on the stigmas twenty-four hours previously, was quite destroyed by that of pollen from a plant belonging to the other form.

. . .

Notes

1   The dimorphism is therefore not a case of 'mere variability'; which, as Darwin said (p. 201), was then the common view.

2   [Darwin's note]   I have proved by numerous experiments . . . that the Polyanthus is a variety of *Primula veris*.

3   Seeds are arranged in seed 'capsules', which are produced in groups of flowers called 'umbels', of which there are several per plant.

4   The numbers in Table 2 are simply those in Table 1, multiplied or divided to give the weight of seeds produced by 10 plants or 100 umbels.

5   So the two forms are not separate sexes. We next consider whether the flowers can self-fertilise.

6   [Darwin's note]   I have shown in my work on the 'Effects of Cross and Self-fertilisation' how greatly the offspring from intercrossed plants profit in height, vigour, and fertility.

Chapter nine

The power of movement in plants (1880)

In the summer of 1877, Darwin began his last important botanical work. He had previously seen, in the climbing plant *Echinocystis lobata*,[1] how 'the uppermost part of each branch is *constantly* and slowly twisting round making a circle in from one and a half to two hours.' The growing plant 'sweeps a circle of from one foot to twenty inches in diameter, and immediately the tendril touches any object its sensitivity causes it immediately to seize it', a pattern of movement that enables the plant to behave in an apparently purposive manner. As 'a clever gardener, my neighbour' had said to Darwin, 'I believe, Sir, the tendrils can see, for wherever I put a plant it finds out any stick near enough.'

Darwin was characteristically modest about the work. He told Hooker 'it is just the sort of niggling work which suits me, and takes up no time and rather rests me whilst writing. . . . The work would hurt my conscience, did I think I could do harder work.' Actually the question of how climbing plants behave in a purposive manner is typically Darwinian, as is his answer to it. In the coral reef theory and in the theory of natural selection too Darwin explains a pattern in nature as the result of a selective retention of a wider range of possibilities. Climbing plants are seen to seek out and climb supporting objects; they do so by moving in many directions and then selectively growing once a support has been found.

213

Climbing plants was published in 1875.[2] Darwin next sought to generalise his theory. He had been struck by the large number of times the climbing habit had evolved independently in plants. He therefore suspected that many – or all – plants must possess the ability to make circular movements in some form, which could be modified into the full climbing habit. 'For in accordance

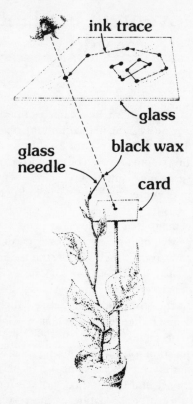

Figure 9.1 Darwin's apparatus to record the movements of plants A glass needle, with a blob of wax at its end, has been glued to the part of the plant under observation; a card, with a black dot on it, has been placed just beneath the needle. An observer could then trace the magnified movements of the plant on a plate of glass above the plant, marking the point where the needle end was above the black dot.

with the principle of evolution it was impossible to account for climbing plants having developed in so many widely different groups, unless all kinds of plants possess powers of movement of an analogous kind.' He set out to discover whether they did. The three gardeners of Down were mobilised, his son Frank enlisted as assistant, and the movements of all parts of as many plant species as possible were recorded. 'This was a tough piece of work.' They made their records with a simple apparatus (Fig. 9.1), which resulted in the kind of diagrammatic trace of movements that can be seen later (e.g. Fig. 9.4). The work gave the result Darwin had hoped for. All plants indeed appeared to undergo constant circular movement – 'circumnutation' as he called it.

But to demonstrate circumnutation was not the only aim of the book. 'I was further led to a rather wide generalisation, viz., that the great and important classes of movement, excited by the light, the attraction of gravity, etc., are all modified forms of the fundamental movement of circumnutation.'

The *Power of movement in plants* accordingly starts with four chapters that demonstrate the circumnutation of different parts of plants; there follow six more, which show how circumnutatory movements can be modified to give rise to 'sleep' movements in leaves, and responses to light and gravity. I have selected Darwin's initial explanation of what 'circumnutation' is, and have then picked on the movements in relation to light, which is the most famous part of the book. Two chapters cover movements in relation to light ('heliotropism'). One shows how heliotropism is a modified form of circumnutation; I have extracted only the short final summary section. The second describes some famous experiments. Heliotropism was then usually thought to be due to the inhibitory action of light on growth. So, if a plant was growing up with light shining on it from one side, the light would inhibit the growth on that side, the opposite side would grow more, and the plant would grow towards the light.

The Darwins exploded this theory, experimentally. They actually experimented on five species, but I have extracted those on only one, the grass *Phalaris*. They observed that the heliotropic plant bends below the tip: they covered the tips of some plants and illuminated them from the side; but the plants did not bend. Evidently heliotropism was not due to the direct action of light, as in the experiment light had been shone directly on the region of the plant that normally bends, and had no effect. Besides the experimental demonstration, Darwin argued that it was unlikely that the direct action of light on growth (even if it existed), could account for the full variety of responses shown by different kinds of plant to light. The reception of light, and the differential growth in relation to it, had to be separate functions. In the theoretical discussion at the end of the chapter (which I have included here) Darwin makes notable use of analogies with the nervous control of purposive behaviour in animals. M. T. Ghiselin has suggested that Darwin was looking for a general theory, which would apply to both plants and animals, of the physiological control of purposive behaviour.

The *Power of movement in plants* was more influential than most of Darwin's books. The book itself, as we have seen, was a collaborative project with his son Francis, and Francis was a capable plant physiologist who would follow the work up. So too did the German physiologist Wilhelm Pfeffer (1845–1920), and his school. In 1931, Darwin's conjecture about the control of heliotropic movements was substantiated with the isolation of the plant growth substance called auxin by the Swedish botanist F. Kögl.

Introduction

The chief object of the present work is to describe and connect together several large classes of movement, common to almost all plants. The most widely prevalent movement is essentially of the same nature as that of the stem of a climbing plant, which bends successively to all points of the compass, so that the tip revolves. This movement has been called by Sachs "revolving nutation;" but we have found it much more convenient to use the terms *circumnutation* and *circumnutate*. As we shall have to say much about this movement, it will be useful here briefly to describe its nature. If we observe a circumnutating stem, which happens at the time to be bent, we will say towards the north, it will be found gradually to bend more and more easterly, until it faces the east; and so onwards to the south, then to the west, and back again to the north. If the movement had been quite regular, the apex would have described a circle, or rather, as the stem is always growing upwards, a circular spiral. But it generally describes irregular elliptical or oval figures; for the apex, after pointing in any one direction, commonly moves back to the opposite side, not, however, returning along the same line. Afterwards other irregular ellipses or ovals are successively described, with their longer axes directed to different points of the compass. Whilst describing such figures, the apex often travels in a zigzag line, or makes small subordinate loops or triangles. In the case of leaves the ellipses are generally narrow.

. . .

In the course of the present volume it will be shown that apparently every growing part of every plant is continually circumnutating, though often on a small scale. Even the stems of seedlings before they have broken through the ground, as well as their buried radicles, circumnutate, as far as the pressure of the surrounding earth permits. In this universally present movement we have the basis or ground-work for the acquirement, according to the requirements of the plant, of the most diversified movements. Thus, the great sweeps made by the stems of twining plants, and by the tendrils of other climbers, result from a mere increase in the amplitude of the ordinary movement of circumnutation. The position which young leaves and other organs ultimately assume is acquired by the circumnutating movement being increased in some one direction. The leaves of various plants are said to sleep at night, and it will be seen that their blades then assume a vertical position through modified circumnutation, in order to protect their upper surfaces from being chilled through radiation. The movements of various organs to the light, which are so general throughout the vegetable kingdom, and occasionally from the light, or transversely with respect to it, are all modified forms of circumnutation; as again are the equally prevalent movements of stems, &c., towards the zenith, and of roots towards the centre of the earth. In accordance with these conclusions, a considerable difficulty in the way of evolution is in part removed, for it might have been asked, how did all these diversified movements for the most different purposes first arise? As the case stands, we know that there is always movement in progress, and its amplitude, or direction, or both, have only to be modified for the good of the plant in relation with internal or external stimuli.

. . . Some other subjects will be discussed. [One of] the two which have interested us most . . . [is] the fact that with some seedling plants the uppermost part alone is sensitive to light, and transmits an influence to the lower part, causing it to bend. If therefore the upper part be wholly protected from light, the lower part may be exposed for hours to it, and yet does not become in the least bent, although this would have occurred quickly if the upper part had been excited by light.

. . .

Heliotropism

Heliotropism.—When a plant which is strongly heliotropic (and species differ much in this respect) is exposed to a bright lateral light, it bends quickly towards it, and the course pursued by the stem is quite or nearly straight. But if the light is much dimmed, or occasionally interrupted, or admitted in only a slightly oblique direction, the course pursued is more or less zigzag; and . . . such zigzag movement results from the elongation or drawing out of the ellipses, loops, &c., which the plant would have described, if it had been illuminated from above.[3] On several occasions we were much struck with this fact, whilst observing the circumnutation of highly sensitive seedlings, which were unintentionally illuminated rather obliquely, or only at successive intervals of time.

For instance, two young seedlings of *Beta vulgaris* were placed in the middle of a room with north-east windows, and were kept covered up, except during each observation which lasted for only a

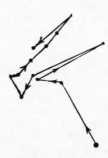

Figure 9.2   Heliotropism of *Beta vulgaris*

Beta vulgaris: circumnutation of hypocotyl, deflected by the light being slightly lateral, traced on a horizontal glass from 8.30 A.M. to 5.30 P.M. Direction of the lighted taper by which it was illuminated, shown by a line joining the first and penultimate dots.

minute or two; but the result was that their hypocotyls bowed themselves to the side, whence some light occasionally entered, in lines which were only slightly zigzag. Although not a single ellipse was even approximately formed, we inferred from the zigzag lines—and, as it proved, correctly—that their hypocotyls were circumnutating, for on the following day these same seedlings were placed in a completely darkened room, and were observed each time by the aid of a small wax taper held almost directly above them, and their movements were traced on a horizontal glass above; and now their hypocotyls clearly circumnutated [Fig. 9.2] . . . yet they moved a short distance towards the side where the taper was held up. If we look at these diagrams, and suppose that the taper had been held more on one side, and that the hypocotyls, still circumnutating, had bent themselves within the same time much more towards the light, long zigzag lines would obviously have been the result.

. . .

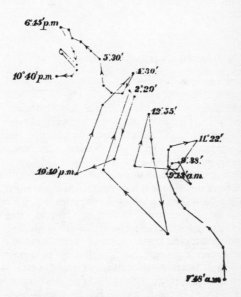

Figure 9.3   Heliotropism of *Tropæolum majus*
Tropæolum majus: heliotropic movement and circumnutation of the epicotyl of a young seedling towards a dull lateral light, traced on a horizontal glass from 7.48 A.M. to 10.40 P.M.

219

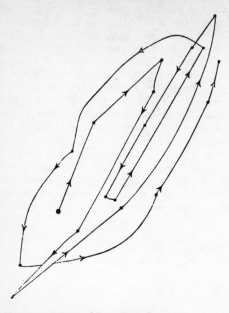

Figure 9.4   Circumnutation of *Brassica oleracea*
Brassica oleracea: ordinary circumnutating movement of the hypocotyl of a seedling plant.

Relation between Circumnutation and Heliotropism.—Any one who will look at the foregoing diagrams, showing the movements of the stems of various plants towards a lateral and more or less dimmed light, will be forced to admit that ordinary circumnutation and heliotropism graduate into one another. When a plant is exposed to a dim lateral light and continues during the whole day bending towards it, receding late in the evening, the movement unquestionably is one of heliotropism. Now, in the case of Tropæolum [Fig. 9.3] the stem or epicotyl obviously circumnutated during the whole day, and yet it continued at the same time to move heliotropically; this latter movement being affected by the apex of each successive elongated figure or ellipse standing nearer to the light than the previous one. . . . The comparison . . . between the ordinary circumnutating movement of a seedling Brassica [Figs 9.4 and 9.5] . . . and their heliotropic movement towards a window protected by blinds [is also instructive]. . . . We have therefore many kinds of gradations from a movement towards the light, which must be considered as one of circumnutation very slightly modified and still consisting of ellipses

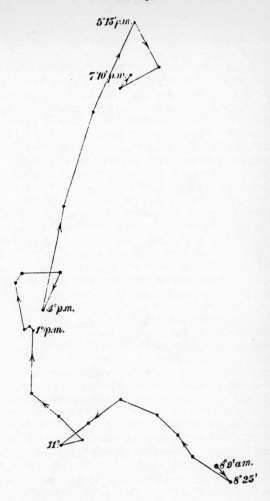

Figure 9.5   Heliotropism of *Brassica oleracea*
Brassica oleracea: heliotropic movement and circumnutation of a hypocotyl towards a very dim lateral light, traced during 11 hours, on a horizontal glass in the morning, and on a vertical glass in the evening.

or circles,—though a movement more or less strongly zigzag, with loops or ellipses occasionally formed,—to a nearly straight, or even quite straight, heliotropic course.

A plant, when exposed to a lateral light, though this may be bright, commonly moves at first in a zigzag line, or even directly from the light; and this no doubt is due to its circumnutating at the time in a direction either opposite to the source of the light, or more or less transversely to it. As soon, however, as the direction of the circumnutating movement nearly coincides with that of the entering light, the plant bends in a straight course towards the light, if this is bright. The course appears to be rendered more and more rapid and rectilinear, in accordance with the degree of brightness of the light—firstly, by the longer axes of the elliptical figures, which the plant continues to describe as long as the light remains very dim, being directed more or less accurately towards its source, and by each successive ellipse being described nearer to the light. Secondly, if the light is only somewhat dimmed, by the acceleration and increase of the movement towards it, and by the retardation or arrestment of that from the light, some lateral movement being still retained, for the light will interfere less with a movement at right angles to its direction, than with one in its own direction. The result is that the course is rendered more or less zigzag and unequal in rate. Lastly, when the light is very bright all lateral movement is lost; and the whole energy of the plant is expended in rendering the circumnutating movement rectilinear and rapid in one direction alone, namely, towards the light.

The common view seems to be that heliotropism is a quite distinct kind of movement from circumnutation; and it may be urged that in the foregoing diagrams we see heliotropism merely combined with, or superimposed on, circumnutation. But if so, it must be assumed that a bright lateral light completely stops circumnutation, for a plant thus exposed moves in a straight line towards it, without describing any ellipses or circles. If the light be somewhat obscured, though amply sufficient to cause the plant to bend towards it, we have more or less plain evidence of still-continued circumnutation. It must further be assumed that it is only a lateral light which has this extraordinary power of stopping circumnutation, for we know that the several plants above experimented on, and all the others which were observed by us whilst growing, continue to circumnutate, however bright the light may be, if it comes from above. Nor should it be forgotten that in the life of each plant, circumnutation precedes heliotropism, for hypocotyls, epicotyls, and petioles circumnutate before they have broken through the ground and have ever felt the influence of light.

222

We are therefore fully justified, as it seems to us, in believing that whenever light enters laterally, it is the movement of circumnutation which gives rise to, or is converted into, heliotropism and apheliotropism.[4] On this view we need not assume against all analogy that a lateral light entirely stops circumnutation; it merely excites the plant to modify its movement for a time in a beneficial manner. The existence of every possible gradation, between a straight course towards a lateral light and a course consisting of a series of loops or ellipses, becomes perfectly intelligible.

. . .

Sensitiveness of plants to light

A pot with seedlings of *Phalaris canariensis*, which had been raised in darkness, was placed in a completely darkened room, at 12 feet from a very small lamp. After 3 h. the cotyledons were doubtfully curved towards the light, and after 7 h. 40 m. from the first exposure, they were all plainly, though slightly, curved towards the lamp. Now, at this distance of 12 feet, the light was so obscure that we could not see the seedlings themselves, nor read the large Roman figures on the white face of a watch, nor see a pencil line on paper, but could just distinguish a line made with Indian ink. It is a more surprising fact that no visible shadow was cast by a pencil held upright on a white card; the seedlings, therefore, were acted on by a difference in the illumination of their two sides, which the human eye could not distinguish. On another occasion even a less degree of light acted, for some cotyledons of *Phalaris* became slightly curved towards the same lamp at a distance of 20 feet; at this distance we could not see a circular dot 2·29 mm. (·09 inch) in diameter made with Indian ink on white paper, though we could just see a dot 3·56 mm. (·14 inch) in diameter; yet a dot of the former size appears large when seen in the light.

We next tried how a small beam of light would act; as this bears on light serving as a guide to seedlings whilst they emerge through fissured or encumbered ground. A pot with seedlings of *Phalaris* was covered by a tin-vessel, having on one side a circular hole 1·23 mm. in diameter (i.e. a little less than the ½₀th of an inch); and the box was placed in front of a paraffin lamp and on another occasion in front of a window; and both times the seedlings were manifestly bent after a few hours towards the little hole.

A more severe trial was now made; little tubes of very thin glass, closed at their upper ends and coated with black varnish, were slipped over the cotyledons of *Phalaris* (which had germinated in darkness) and just fitted them. Narrow stripes of the varnish had been previously scraped off one side, through which alone light could enter; and their dimensions were afterwards measured under the microscope. As a control experiment, similar unvarnished and transparent tubes were tried, and they did not prevent the cotyledons bending towards the light. Two cotyledons were placed before a south-west window, one of which was illuminated by a stripe in the varnish, only ·004 inch (0·1 mm.) in breadth and ·016 inch (0·4 mm.) in length; and the other by a stripe ·008 inch in breadth and ·06 inch in length. The seedlings were examined after an exposure of 7 h. 40 m., and were found to be manifestly bowed towards the light. Some other cotyledons were at the same time treated similarly, excepting that the little stripes were directed not to the sky, but in such a manner that they received only the diffused light from the room; and these cotyledons did not become at all bowed. Seven other cotyledons were illuminated through narrow, but comparatively long, cleared stripes in the varnish—namely, in breadth between ·01 and ·026 inch, and in length between ·15 and ·3 inch; and these all became bowed to the side, by which light entered through the stripes, whether these were directed towards the sky or to one side of the room. That light passing through a hole only ·004 inch in breadth by ·016 in length, should induce curvature, seems to us a surprising fact.

. . .

The cotyledons of *Phalaris* bend much more slowly towards a very obscure light than towards a bright one. Thus, in the experiments with seedlings placed in a dark room at 12 feet from a very small lamp, they were just perceptibly and doubtfully curved towards it after 3 h., and only slightly, yet certainly, after 4 h. After 8 h. 40 m. the chords of their arcs were deflected from the perpendicular by an average angle of only 16°. Had the light been bright, they would have become much more curved in between 1 and 2 h. Several trials were made with seedlings placed at various distances from a small lamp in a dark room; but we will give only one trial. Six pots were placed at distances of 2, 4, 8, 12, 16, and 20 feet from the lamp, before which they were left for 4 h. As light decreases in a geometrical ratio, the seedlings in the 2nd pot received ¼th, those in the 3rd pot ¹⁄₁₆th, those in the 4th ¹⁄₃₆th, those in the 5th ¹⁄₆₄th, and those in the 6th ¹⁄₁₀₀th of the light received by the seedlings in the first or nearest pot. Therefore it might have been expected that there would have been an immense difference in the degree of their heliotropic curvature in the several pots; and there was a well-marked difference between those which stood nearest and furthest from the lamp, but the difference in each successive pair of pots was extremely small. In order to avoid prejudice, we asked three persons, who knew nothing about the experiment, to arrange the pots in order according to the degree of curvature of the cotyledons. The first person arranged them in proper order, but doubted long between the 12 feet and 16 feet pots; yet these two received light in the proportion of 36 to 64. The second person also arranged them properly, but doubted between the 8 feet and 12 feet pots, which received light in the proportion of 16 to 36. The third person arranged them in wrong order, and doubted about four of the pots. This evidence shows conclusively how little the curvature of the seedlings differed in the successive pots, in comparison with the great difference in the amount of light which they received; and it should be noted that there was no excess of superfluous light, for the cotyledons became but little and slowly curved even in the nearest pot. Close to the 6th pot, at the distance of 20 feet from the lamp, the light allowed us just to distinguish a dot 3·56 mm. (·14 inch) in diameter, made with Indian ink on white paper, but not a dot 2·29 mm. (0·9 inch) in diameter.

The degree of curvature of the cotyledons of *Phalaris* within a given time, depends not merely on the amount of lateral light which they may then receive, but on that which they have previously received from above and on all sides.... Of two pots containing seedlings of *Phalaris* which had germinated in darkness, one was still kept in the dark, and the other was exposed (Sept. 26th) to the light in a greenhouse during a cloudy day and on the following bright morning. On this morning (27th), at 10.30 A.M., both pots were placed in a box, blackened within and open in front, before a north-east window, protected by a linen and muslin blind and by a towel, so that but little light was admitted, though the sky was bright. Whenever the pots were looked at, this was done as quickly as possible, and the cotyledons were then held transversely with respect to the light, so that their curvature could not have been thus increased or diminished. After 50 m. the seedlings which had previously been kept in darkness, were perhaps, and after 70 m. were certainly, though very slightly, towards the window. After 85 m. some of the seedlings, which had previously been illuminated, were perhaps a little affected, and after 100 m. some of the younger ones were certainly a little curved towards the light. At this time (i.e. after 100 m.) there was a plain difference in the curvature of the seedlings in the two pots. After 2 h. 12 m. the chords of the arcs of four of the most strongly curved seedlings in each pot were measured, and the mean angle from the perpendicular of those which had previously been kept in darkness was 19°, and of those which had previously been illuminated only 7°.

. . .

In order to observe how long the after-effects of light lasted, a pot with seedlings of *Phalaris*, which had germinated in darkness, was placed at 10.40 A.M. before a north-east window, being protected on all other sides from the light; and the movement of a cotyledon was traced on a horizontal glass. It circumnutated about the same space for the first 24 m., and during the next 1 h. 33 m. moved rapidly towards the light. The light was now (i.e. after 1 h. 57 m.) completely excluded, but the cotyledon continued bending in the same direction as before, certainly for more than 15 m., probably for about 27 m. The doubt arose from the necessity of not looking at the seedlings often, and thus exposing them, though momentarily, to the light. This same seedling was now kept in the dark, until 2.18 P.M., by which time it had reacquired . . . its original upright position, when it was again exposed to the light from a clouded sky. By 3 P.M., it had moved a very short distance towards the light, but during the next 45 m. travelled quickly towards it. After this exposure of 1 h. 27 m. to a rather dull sky, the light was again completely excluded, but the cotyledon continued to bend in the same direction as before for 14 m. within a very small limit of error. It was then placed in the dark, and it now moved backwards, so that after 1 h. 7 m. it stood close to where it had started from at 2.18 P.M. These observations show that the cotyledons of *Phalaris*, after being exposed to a lateral light, continue to bend in the same direction for between a quarter and half an hour.

. . .

In our various experiments we were often struck with the accuracy with which seedlings pointed to a light although of small size. To test this, many seedlings of *Phalaris*, which had germinated in darkness in a very narrow box several feet in length, were placed in a darkened room near to and in front of a lamp having a small cylindrical wick. The cotyledons at the two ends and in the central part of the box, would therefore have to bend in widely different directions in order to point to the light. After they had become rectangularly bent, a long white thread was stretched by two persons, close over and parallel, first to one and then to another cotyledon; and the thread was found in almost very case actually to intersect the small circular wick of the now extinguished lamp. The deviation from accuracy never exceeded, as far as we could judge, a degree or two. This extreme accuracy seems at first surprising, but is not really so, for an upright cylindrical stem, whatever its position may be with respect to the light, would have exactly half its circumference illuminated and half in shadow; and as the difference in illumination of the two sides is the

226

exciting cause of heliotropism, a cylinder would naturally bend with much accuracy towards the light. The cotyledons, however, of *Phalaris* are not cylindrical, but oval in section; and the longer axis was to the shorter axis (in the one which was measured) as 100 to 70. Nevertheless, no difference could be detected in the accuracy of their bending, whether they stood with their broad or narrow sides facing the light, or in any intermediate position; and so it was with the cotyledons of *Avena sativa*, which are likewise oval in section. Now, a little reflection will show that in whatever position the cotyledons may stand, there will be a line of greatest illumination, exactly fronting the light, and on each side of this line an equal amount of light will be received; but if the oval stands obliquely with respect to the light, this will be diffused over a wider surface on one side of the central line than on the other. We may therefore infer that the same amount of light, whether diffused over a wider surface or concentrated on a smaller surface, produces exactly the same effect; for the cotyledons in the long narrow box stood in all sorts of positions with reference to the light, yet all pointed truly towards it.

That the bending of the cotyledons to the light depends on the illumination of one whole side or on the obscuration of the whole opposite side, and not on a narrow longitudinal zone in the line of the light being affected, was shown by the effects of painting longitudinally with Indian ink one side of five cotyledons of *Phalaris*. These were then placed on a table near to a south-west window, and the painted half was directed either to the right or left. The result was that instead of bending in a direct line towards the window, they were deflected from the window and towards the unpainted side, by the following angles, 35°, 83°, 31°, 43°, and 39°. It should be remarked that it was hardly possible to paint one-half accurately, or to place all the seedlings which are oval in section in quite the same position relatively to the light; and this will account for the differences in the angles. Five cotyledons of *Avena* were also painted in the same manner, but with greater care; and they were laterally deflected from the line of the window, towards the unpainted side, by the following angles, 44°, 44°, 55°, 51°, and 57°. This deflection of the cotyledons from the window is intelligible, for the whole unpainted side must have received some light, whereas the opposite and painted side received none; but a narrow zone on the unpainted side directly in front of the window will have received most light, and all the hinder parts (half an oval in section) less and less light in varying degrees; and we may conclude that the angle of deflection is the resultant of the action of the light over the whole of the unpainted side.

It should have been premised that painting with Indian ink does not injure plants, at least within several hours; and it could injure them only by stopping respiration. To ascertain whether injury was thus soon caused, the upper halves of 8 cotyledons of *Avena* were thickly coated with transparent matter,—4 with gum, and 4 with gelatine; they were placed in the morning before a window, and by the evening they were normally bowed towards the light, although the coating now consisted of dry crusts of gum and gelatine. . . .

Localised sensitiveness to light, and its transmitted effects

Phalaris canariensis.—Whilst observing the accuracy with which the cotyledons of this plant became bent towards the light of a small lamp, we were impressed with the idea that the uppermost part determined the direction of the curvature of the lower part. When the cotyledons are exposed to a lateral light, the upper part bends first, and afterwards the bending gradually extends down to the base, and, as we shall presently see, even a little beneath the ground. This holds good with cotyledons from less than ·1 inch (one was observed to act in this manner which was only ·03 in height) to about ·5 of an inch in height; but when they have grown to nearly an inch in height, the basal part, for a length of ·15 to ·2 of an inch above the ground, ceases to bend. As with young cotyledons the lower part goes on bending, after the upper part has become well arched towards a lateral light, the apex would ultimately point to the ground instead of to the light, did not the upper part reverse its curvature and straighten itself, as soon as the upper convex surface of the bowed-down portion received more light than the lower concave surface. The position ultimately assumed by young and upright cotyledons, exposed to light entering obliquely from above through a window, is shown in the accompanying figure [Fig. 9.6]; and here it may be seen that the whole upper part has become very nearly straight. When the cotyledons were exposed before a bright lamp, standing on the same level with them, the upper part, which was at first greatly arched towards the light, became straight and strictly parallel with the surface of the soil in the pots; the basal part being now rectangularly bent. All this great amount of curvature, together with the subsequent straightening of the upper part, was often effected in a few hours.

Figure 9.6   Heliotropism of *Phalaris*
Phalaris canariensis: cotyledons after exposure in a box open on one side in front of a south-west window during 8 h. Curvature towards the light accurately traced. The short horizontal lines show the level of the ground.

After the uppermost part has become bowed a little to the light, its overhanging weight must tend to increase the curvature of the lower part; but any such effect was shown in several ways to be quite insignificant. When little caps of tin-foil (hereafter to be described) were placed on the summits of the cotyledons, though this must have added considerably to their weight, the rate or amount of bending was not thus increased. But the best evidence was afforded by placing pots with seedlings of *Phalaris* before a lamp in such a position, that the cotyledons were horizontally extended and projected at right angles to the line of light. In the course of $5\frac{1}{2}$ h. they were directed towards the light with their bases bent at right angles; and this abrupt curvature could not have been aided in the least by the weight of the upper part, which acted at right angles to the plane of curvature.

It will be shown that when the upper halves of the cotyledons of *Phalaris* and *Avena* were enclosed in little pipes of tin-foil or blackened glass, in which case the upper part was mechanically prevented from bending, the lower and unenclosed part did not bend when exposed to a lateral light; and it occurred to us that this fact might be due, not to the exclusion of the light from the upper part, but to some necessity of the bending gradually travelling down the cotyledons, so that unless the upper part first became bent, the lower could not bend, however much it might be stimulated. It was necessary for our purpose to ascertain whether this notion was true, and it was proved false; for the lower halves of several cotyledons became bowed to the light, although their upper halves were enclosed in little glass tubes (not blackened), which prevented, as far as we could judge, their bending. Nevertheless, as the part within the tube might possibly bend a very little, fine rigid rods or flat splinters of thin glass were cemented with shellac to one side of the upper part of 15 cotyledons; and in six cases they were in addition tied on with threads. They were thus forced to remain quite straight. The result was that the lower halves of all became bowed to the light, but generally not in so great a degree as the corresponding part of the free seedlings in the same pots; and this may perhaps be accounted for by some slight degree of injury having been caused by a considerable surface having been smeared with shellac. . . .

To test our belief that the upper part of the cotyledons of *Phalaris*, when exposed to a lateral light, regulates the bending of the lower part, many experiments were tried; but most of our first attempts proved useless from various causes not worth specifying. Seven cotyledons had their tips cut off for lengths varying between ·1 and ·16 of an inch, and these, when left exposed all day to a lateral light, remained upright. In another set of 7 cotyledons, the tips were cut off for a length of only about ·05 of an inch (1·27 mm.) and these became bowed towards a lateral light, but not nearly so much as the many other seedlings in the same pots. This latter case shows that cutting off the tips does not by itself injure the plants so seriously as to prevent heliotropism; but we thought at the time, that such injury might follow when a greater length was cut off, as in the first set of experiments. . . .

We next tried the effects of covering the upper part of the cotyledons of *Phalaris* with little caps which were impermeable to light; the whole lower part being left fully exposed before a south-west window or a bright paraffin lamp. Some of the caps were made of extremely thin tin-foil blackened within; these had the disadvantage of occasionally, though rarely, being too heavy, especially when twice folded. The basal edges could be pressed into close contact with the cotyledons; though this again required care to prevent injuring them. Nevertheless, any injury thus caused could be detected by removing the caps, and trying whether the cotyledons were then sensitive to light. Other caps were made of tubes of the thinnest glass, which when painted black served well, with the one great disadvantage that the lower ends could not be closed. But tubes were used which fitted the cotyledons almost closely, and black paper was placed on the soil round each, to check the upward reflection of light from the soil. Such tubes were in one respect far better than caps of tin-foil, as it was possible to cover at the same time some cotyledons with transparent and others with opaque tubes; and thus our experiments could be controlled. It should be kept in mind that young cotyledons were selected for trial, and that these when not interfered with become bowed down to the ground towards the light.

We will begin with the glass-tubes. The summits of nine cotyledons, differing somewhat in height, were enclosed for rather less than half their lengths in uncoloured or transparent tubes; and these were then exposed before a south-west window on a bright day for 8 h. All of them became strongly curved towards the light, in the same degree as the many other free seedlings in the same pots; so that the glass-tubes certainly did not prevent the cotyledons from bending towards the light. Nineteen other cotyledons were, at the same time, similarly enclosed in tubes thickly painted with Indian ink. On five of them, the paint, to our surprise, contracted after exposure to the sunlight, and very narrow cracks were formed, through which a little light entered; and these five cases were rejected. Of the remaining 14 cotyledons, the lower halves of which had been fully exposed to the light for the whole time, 7 continued quite straight and upright; 1 was considerably bowed to the light, and 6 were slightly bowed, but with the exposed bases of most of them almost or

230

quite straight. It is possible that some light may have been reflected upwards from the soil and entered the bases of these 7 tubes, as the sun shone brightly, though bits of blackened paper had been placed on the soil round them. Nevertheless, the 7 cotyledons which were slightly bowed, together with the 7 upright ones, presented a most remarkable contrast in appearance with the many other seedlings in the same pots to which nothing had been done. The blackened tubes were then removed from 10 of these seedlings, and they were now exposed before a lamp for 8 h.: 9 of them became greatly, and 1 moderately, curved towards the light, proving that the previous absence of any curvature in the basal part, or the presence of only a slight degree of curvature there, was due to the exclusion of light from the upper part.

Similar observations were made on 12 younger cotyledons with their upper halves enclosed within glass-tubes coated with black varnish, and with their lower halves fully exposed to bright sunshine. In these younger seedlings, the sensitive zone seems to extend rather lower down, as was observed on some other occasions, for two became almost as much curved towards the light as the free seedlings; and the remaining ten were slightly curved, although the basal part of several of them, which normally becomes more curved than any other part, exhibited hardly a trace of curvature. These 12 seedlings taken together differed greatly in their degree of curvature from all the many other seedlings in the same pots.

Better evidence of the efficiency of the blackened tubes was incidentally afforded by some experiments hereafter to be given, in which the upper halves of 14 cotyledons were enclosed in tubes from which an extremely narrow stripe of the black varnish had been scraped off. These cleared stripes were not directed towards the window, but obliquely to one side of the room, so that only a very little light could act on the upper halves of the cotyledons. These 14 seedlings remained during eight hours of exposure before a south-west window on a hazy day quite upright; whereas all the other many free seedlings in the same pots became greatly bowed towards the light.

We will not turn to the trials with caps made of very thin tin-foil. These were placed at different times on the summits of 24 cotyledons, and they extended down for a length of between ·15 and ·2 of an inch. The seedlings were exposed to a lateral light for periods varying between 6 h. 30 m. and 7 h. 45 m., which sufficed to cause all the other seedlings in the same pots to become almost rectangularly bent towards the light. They varied in height from only ·04 to 1·15 inch, but the greater number were about ·75 inch. Of the 24 cotyledons with their summits thus protected, 3 became much bent, but not in the direction of the light, and as they did not straighten themselves through apogeotropism during the following night, either the caps were too heavy or the plants themselves were in a weak condition; and these three cases may be excluded. There are left for consideration 21 cotyledons; of these 17 remained all the time quite upright; the other 4 became slightly inclined to the light, but not in a degree comparable with that of the many free seedlings in the same pots. As the glass-tubes, when unpainted, did not prevent the cotyledons from becoming greatly bowed, it cannot be supposed that the caps of very thin tin-foil did so, except through the exclusion of the light. To prove that the plants had not been injured, the caps were removed from 6 of the upright seedlings, and these were exposed before a paraffin lamp for the same length of time as before, and they now all became greatly curved towards the light.

As caps between ·15 and ·2 of an inch in depth were thus proved to be highly efficient in preventing the cotyledons from bending towards the light, 8 other cotyledons were protected with caps between only ·06 and ·12 in depth. Of these, two remained vertical, one was considerably and five slightly curved towards the light, but far less so than the free seedlings in the same pots.

Another trial was made in a different manner, namely, by bandaging with strips of tin-foil, about ·2 in breadth, the upper part, but not the actual summit, of eight moderately young seedlings a little over half an inch in height. The summits and the basal parts were thus left fully exposed to a lateral light during 8 h.; an upper intermediate zone being protected. With four of these seedlings the summits were exposed for a length of ·05 inch, and in two of them this part became curved towards the light, but the whole lower part remained quite upright; whereas the entire length of the other two seedlings became slightly curved towards the light. The summits of the four other seedlings were exposed for a length of ·04 inch, and of these one remained almost upright, whilst the other three became considerably curved towards the light. The many free seedlings in the same pots were all greatly curved towards the light.

From these several sets of experiments, including those with the glass-tubes, and those when the tips were cut off, we may infer that the exclusion of light from the upper part of the cotyledons of *Phalaris* prevents the lower part, though fully exposed to a lateral light, from becoming curved. The summit for a length of ·04 or ·05 of an inch, though it is itself sensitive and curves towards the light, has only a slight power of causing the lower part to bend. Nor has the exclusion of light from the summit for a length of ·1 of an inch a strong influence on the curvature of the lower part. On the other hand, an exclusion for a length of between ·15 and ·2 of an inch, or of the whole

232

upper half, plainly prevents the lower and fully illuminated part from becoming curved in the manner [see Fig. 9.6] which invariably occurs when a free cotyledon is exposed to a lateral light. With very young seedlings the sensitive zone seems to extend rather lower down relatively to their height than in older seedlings. We must therefore conclude that when seedlings are freely exposed to a lateral light some influence is transmitted from the upper to the lower part, causing the latter to bend.

This conclusion is supported by what may be seen to occur on a small scale, especially with young cotyledons, without any artificial exclusion of the light; for they bend beneath the earth where no light can enter.

. . .

Concluding remarks

We do not know whether it is a general rule with seedling plants that the illumination of the upper part determines the curvature of the lower part. But as this occurred in the four[5] species examined by us, belonging to such distinct families as the Gramineæ, Cruciferæ, and Chenopodeæ, it is probably of common occurrence. It can hardly fail to be of service to seedlings, by aiding them to find the shortest path from the buried seed to the light, on nearly the same principle that the eyes of most of the lower crawling animals are seated at the anterior ends of their bodies. It is extremely doubtful whether with fully developed plants the illumination of one part ever affects the curvature of another part. The summits of 5 young plants of *Asparagus officinalis* (varying in height between 1·1 and 2·7 inches, and consisting of several short internodes) were covered with caps of tin-foil from 0·3 to 0·35 inch in depth; and the lower uncovered parts became as much curved towards a lateral light, as were the free seedlings in the same pots. Other seedlings of the same plant had their summits painted with Indian ink with the same negative result. Pieces of blackened paper were gummed to the edges and over the blades of some leaves on young plants of *Tropæolum majus* and *Ranunculus ficaria*; these were then placed in a box before a window, and the petioles of the protected leaves became curved towards the light, as much as those of the unprotected leaves.

The foregoing cases with respect to seedling plants have been fully described, not only because the transmission of any effect from light is a new physiological fact, but because we think it tends to modify somewhat the current views on heliotropic movements. Until lately such movements were believed to result simply from increased growth on the shaded side. . . . All observers apparently believe that

233

light acts directly on the part which bends, but we have seen with the above described seedlings that this is not the case. Their lower halves were brightly illuminated for hours, and yet did not bend in the least towards the light, though this is the part which under ordinary circumstances bends the most. It is a still more striking fact, that the faint illumination of a narrow stripe on one side of the upper part of the cotyledons of *Phalaris* determined the direction of the curvature of the lower part; so that this latter part did not bend towards the bright light by which it had been fully illuminated, but obliquely towards one side where only a little light entered. These results seem to imply the presence of some matter in the upper part which is acted on by light, and which transmits its effects to the lower part. It has been shown that this transmission is independent of the bending of the upper sensitive part. . . .

Light exerts a powerful influence on most vegetable tissues, and there can be no doubt that it generally tends to check their growth. But when the two sides of a plant are illuminated in a slightly different degree, it does not necessarily follow that the bending towards the illuminated side is caused by changes in the tissues of the same nature as those which lead to increased growth in darkness. We know at least that a part may bend from the light, and yet its growth may not be favoured by light. This is the case with the radicles of *Sinapis alba*, which are plainly apheliotropic; nevertheless, they grow quicker in darkness than in light. So it is with many aërial roots, according to Wiesner; but there are other opposed cases. It appears, therefore, that light does not determine the growth of apheliotropic parts in any uniform manner.

We should bear in mind that the power of bending to the light is highly beneficial to most plants. There is therefore no improbability in this power having been specially acquired. In several respects light seems to act on plants in nearly the same manner as it does on animals by means of the nervous system. With seedlings the effect, as we have just seen, is transmitted from one part to another. An animal may be excited to move by a very small amount of light; and it has been shown that a difference in the illumination of the two sides of the cotyledons of *Phalaris*, which could not be distinguished by the human eye, sufficed to cause them to bend. It has also been shown that there is no close parallelism between the amount of light which acts on a plant and its degree of curvature; it was indeed hardly possible to perceive any difference in the curvature of some seedlings of *Phalaris* exposed to a light, which, though dim, was very much brighter than that to which others had been exposed. The retina, after being stimulated by a bright light, feels the effect for some time; and *Phalaris* continued to bend for nearly half an hour towards the side which had been illuminated. The retina cannot perceive a dim light after it has been exposed to a bright one; and plants which had been kept in the daylight during the previous day and morning, did not move so soon towards an obscure lateral light as did others which had been kept in complete darkness.

. . .

Even in the case of ordinary heliotropic movements, it is hardly credible that they result directly from the action of the light, without any special adaptation. We may illustrate what we mean by the hygroscopic movements of plants: if the tissues on one side of an organ permit of rapid evaporation, they will dry quickly and contract, causing the part to bend to this side. Now the wonderfully complex movements of the pollinia of *Orchis pyramidalis*, by which they clasp the proboscis of a moth and afterwards change their position for the sake of depositing the pollen-masses on the double stigma—or again the twisting movements, by which certain seeds bury themselves in the ground—follow from the manner of drying of the parts in question; yet no one will suppose that these results have been gained without special adaptation. Similarly, we are led to believe in adaptation when we see the hypocotyl of a seedling, which contains chlorophyll, bending to the light; for although it thus receives less light, being now shaded by its own cotyledons, it places them—the more important organs—in the best position to be fully illuminated. The hypocotyl may therefore be said to sacrifice itself for the good of the cotyledons, or rather of the whole plant. But if it be prevented from bending, as must sometimes occur with seedlings springing up in an entangled mass of vegetation, the cotyledons themselves bend so as to face the light; the one farthest off rising up, and that nearest to the light sinking down, or both twisting laterally. We may, also, suspect that the extreme sensitiviness to light of the upper part of the sheath-like cotyledons of the Gramineæ, and their power of transmitting its effects to the lower part, are specialised arrangements for finding the shortest path to the light. With plants growing on a bank, or thrown prostrate by the wind, the manner in which the leaves move, even rotating on their own axes, so that their upper surfaces may be again directed to the light, is a striking phenomenon. Such facts are rendered more striking when we remember that too intense a light injures the chlorophyll, and that the leaflets of several Leguminosæ when thus exposed bend upwards and present their edges towards the sun, thus escaping injury. On the other hand, the leaflets of *Averrhoa* and *Oxalis*, when similarly exposed, bend downwards.

It was shown in the last chapter that heliotropism is a modified form of circumnutation; and as every growing part of every plant circumnutates more or less, we can understand how it is that the power of bending to the light has been acquired by such a multitude of plants throughout the vegetable kingdom. The manner in which a circumnutating movement—that is, one consisting of a succession of irregular ellipses or loops—is gradually converted into a rectilinear course towards the light, has been already explained. First, we have a succession of ellipses with their longer axes directed towards the light, each of which is described nearer and nearer to its source; then the loops are drawn out into a strongly pronounced zigzag line, with here and there a small loop still formed. At the same time that the movement towards the light is increased in extent and accelerated, that in the opposite direction is lessened and retarded, and at last stopped. The zigzag movement to either side is likewise gradually lessened, so that finally the course becomes rectilinear. Thus under the stimulus of a fairly bright light there is no useless expenditure of force.

As with plants every character is more or less variable, there seems to be no great difficulty in believing that their circumnutating movements may have been increased or modified in any beneficial manner by the preservation of varying individuals. The inheritance of habitual movements is a necessary contingent for this process of selection, or the survival of the fittest; and we have seen good reason to believe that habitual movements are inherited by plants. In the case of twining species the circumnutating movements have been increased in amplitude and rendered more circular; the stimulus being here an internal or innate one. With sleeping plants the movements have been increased in amplitude and often changed in direction; and here the stimulus is the alternation of light and darkness, aided, however, by inheritance. In the case of heliotropism, the stimulus is the unequal illumination of the two sides of the plant, and this determines, as in the foregoing cases, the modification of the circumnutating movement in such a manner that the organ bends to the light.

. . .

Notes

1 A North American wild cucumber; the seeds had been sent by Asa Gray.

2 It had originally been published as a long paper in 1865; the book was a revised form.

3 Which would have been the normal circumnutatory movements.

4 Apheliotropism is growth *away* from light, or negative heliotropism.

5 Only one is represented here.

236

---------- *Chapter ten* ----------

The formation of vegetable mould, through the action of worms (1881)

The book on worms was Darwin's last, but the subject was a long-standing interest. He described it as the 'completion of a short paper read before the Geological Society more than 40 years ago.' It has several themes; the main one is the way in which such an insensibly slight process as the bringing of soil to the surface by earthworms can, in its cumulative effect over a long period, be of substantial geological importance. He measured the rate at which worms brought soil to the surface by watching, over the years, stones sink into the ground. One such measurement was begun soon after he moved to Down.

On 20 December 1842 'a quantity of chalk was spread over a part of a field near my house. The chalk was laid on the land for the sake of observing at some future period to what depth it would become buried. At the end of November, 1871, that is after an interval of 29 years, a trench was dug across this part of the field; and a line of white nodules could be traced on both sides of the trench, at a depth of about 7 inches from the surface. The mould, therefore, (excluding the turf) had here been thrown up at an average rate of .22 inches per year.'

I have extracted below some of Darwin's calculations on the

rate of formation of vegetable mould. The burial of ancient monuments is a striking illustration of the action of worms ('archaeologists ought to be grateful to worms'); and I have extracted the most thorough example, the Roman town of Silchester. The book also contains an important chapter of observations on the habits of worms. Darwin's experiments were designed to find out the 'mental powers' of worms. For instance, he observed that, when worms pull leaves into their burrows, they seize the leaf (in the mouth) so that it can be dragged most easily. He put out various kinds of leaves, some experimentally altered, and cut out paper 'leaves', to see whether worms would also grip them in such a way as to drag them easily. Using a paper leaf of the shape of an isoceles triangle, for example, he counted in what proportions the different parts of the triangle were gripped. The worms consistently seized the appropriate part, the tip. 'We may therefore infer – improbable as is the inference – that worms are able by some means to judge which is the best end to draw triangles of paper into their burrows.' Delightful as these experiments are, I have not included an extract, as they are not the essential theme of the book.

Darwin began the *Formation of vegetable mould* in 1880, before the *Power of movement* was completely off his hands. 'As far as I can judge it will be a curious little book', he told Victor Carus, though 'I have perhaps treated it in foolish detail.' It is the detail that gives the book its charm. For it is a charming book, perhaps the most pleasant to read of all his works. It was a great success, and sold at a higher rate than even the *Origin*. 'It has been a complete surprise to me how many people have cared for the subject', Darwin wrote: but it was his own delightful treatment, as much as the subject itself, that made the book a success. Now that the worms of Down House have raised up another 22 inches of vegetable mould, it is still a pleasure to be introduced to them.

Introduction

The share which worms have taken in the formation of the layer of vegetable mould, which covers the whole surface of the land in every moderately humid country, is the subject of the present volume. This mould is generally of a blackish colour and a few inches in thickness. In different districts it differs but little in appearance, although it may rest on various subsoils. The uniform fineness of the particles of which it is composed is one of its chief characteristic features; and this may be well observed in any gravelly country, where a recently-ploughed field immediately adjoins one which has long remained undisturbed for pasture, and where the vegetable mould is exposed on the sides of a ditch or hole. The subject may appear an insignificant one, but we shall see that it possesses some interest . . .

As I was led to keep in my study during many months worms in pots filled with earth, I became interested in them, and wished to learn how far they acted consciously, and how much mental power they displayed. I was the more desirous to learn something on this head, as few observations of this kind have been made, as far as I know, on animals so low in the scale of organization and so poorly provided with sense-organs, as are earth-worms.

In the year 1837, a short paper was read by me before the Geological Society of London, "On the Formation of Mould", in which it was shown that small fragments of burnt marl, cinders, &c., which had been thickly strewed over the surface of several meadows, were found after a few years lying at the depth of some inches beneath the turf but still forming a layer. This apparent sinking of superficial bodies is due, as was first suggested to me by Mr. Wedgwood of Maer Hall in Staffordshire, to the large quantity of fine earth continually brought up to the surface by worms in the form of castings. These castings are sooner or later spread out and cover up any object left on the surface. I was thus led to conclude that all the vegetable mould over the whole country has passed many times through, and will again pass many times through, the intestinal canals of worms. Hence the term "animal mould" would be in some respects more appropriate than that commonly used of "vegetable mould".

. . .

The amount of fine earth brought up by worms to the surface

. . . We now come to the more immediate subject of this volume, namely the amount of earth which is brought up by worms from beneath the surface, and is afterwards spread out more or less completely by the rain and wind. The amount can be judged of by two methods,—by the rate at which objects left on the surface are buried, and more accurately by weighing the quantity brought up within a given time. We will begin with the first method, as it was first followed.

Near Maer Hall in Staffordshire, quick-lime had been spread about the year 1827 thickly over a field of good pasture-land, which had not since been ploughed. Some square holes were dug in this field in the beginning of October 1837; and the sections showed a layer of turf, formed by the matted roots of the grasses, ½ inch in thickness, beneath which, at a depth of 2½ inches (or 3 inches from the surface), a layer of the lime in powder or in small lumps could be distinctly seen running all round the vertical sides of the holes. The soil beneath the layer of lime was either gravelly or of a coarse sandy nature, and differed considerably in appearance from the overlying dark-coloured fine mould. Coal-cinders had been spread over a part of this same field either in the year 1833 or 1834; and when the above holes were dug, that is after an interval of 3 or 4 years, the cinders formed a line of black spots round the holes, at a depth of 1 inch beneath the surface, parallel to and above the white layer of lime. Over another part of this field cinders had been strewed, only about half-a-year before, and these either still lay on the surface or were entangled among the roots of the grasses; and I here saw the commencement of the burying process, for worm-castings had been heaped on several of the smaller fragments. After an interval of 4¾ years this field was re-examined, and now the two layers of lime and cinders were found almost everywhere at a greater depth than before by nearly 1 inch, we will say by ¾ of an inch. Therefore mould to an average thickness of ·22 of an inch had been annually brought up by the worms, and had been spread over the surface of this field.

. . .

At Stonehenge, some of the outer Druidical stones are now prostrate, having fallen at a remote but unknown period; and these have become buried to a moderate depth in the ground. They are surrounded by sloping borders of turf, on which recent castings were seen. Close to one of these fallen stones, which was 17 ft. long, 6 ft. broad, and 28½ inches thick, a hole was dug; and here the vegetable mould was at least 9½ inches in thickness. At this depth a flint was found, and a little higher up on one side of the hole a fragment of glass. The base of the stone lay about 9½ inches beneath the level of the surrounding ground, and its upper surface 19 inches above the ground.

A hole was also dug close to a second huge stone, which in falling had broken into two pieces; and this must have happened long ago, judging from the weathered aspect of the fractured ends. The base was buried to a depth of 10 inches, as was ascertained by driving an iron skewer horizontally into the ground beneath it. The vegetable mould forming the turf-covered sloping border round the stone, on which many castings had recently been ejected, was 10 inches in

thickness; and most of this mould must have been brought up by
worms from beneath its base. At a distance of 8 yards from the stone,
the mould was only 5½ inches in thickness (with a piece of tobacco
pipe at a depth of 4 inches), and this rested on broken flint and chalk
which could not have easily yielded to the pressure or weight of the
stone.

A straight rod was fixed horizontally (by the aid of a spirit-level)
across a third fallen stone, which was 7 feet 9 inches long; and the
contour of the projecting parts and of the adjoining ground, which
was not quite level, was thus ascertained, as shown in the accompanying diagram [Fig. 10.1] on a scale of ½ inch to a foot. The turf-covered
border sloped up to the stone on one side to a height of 4 inches, and
on the opposite side to only 2½ inches above the general level. A
hole was dug on the eastern side, and the base of the stone was here
found to lie at a depth of 4 inches beneath the general level of the
ground, and of 8 inches beneath the top of the sloping turf-covered
border.

. . .

*Weight of the earth ejected from a single burrow, and from all the
burrows within a given space.*

. . .

A lady, on whose accuracy I can implicitly rely, offered to collect
during a year all the castings thrown up on two separate square yards,
near Leith Hill Place, in Surrey.[1] The amount collected was, however,
somewhat less than that originally ejected by the worms; for as I have
repeatedly observed a good deal of the finest earth is washed away,
whenever castings are thrown up during or shortly before heavy rain.
Small portions also adhered to the surrounding blades of grass, and it
required too much time to detach every one of them. On sandy soil,

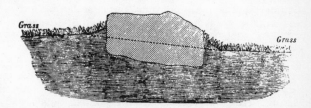

Figure 10.1   A fallen druidical stone at Stonehenge
Section through one of the fallen Druidical stones at Stonehenge, showing how
much it had sunk into the ground. Scale ½ inch to 1 foot.

as in the present instance, castings are liable to crumble after dry weather, and particles were thus often lost. The lady also occasionally left home for a week or two, and at such times the castings must have suffered still greater loss from exposure to the weather. These losses were, however, compensated to some extent by the collections having been made on one of the squares for four days, and on the other square for two days more than the year.

A space was selected (October 9th, 1870) on a broad, grass-covered terrace, which had been mowed and swept during many years. It faced the south, but was shaded during part of the day by trees. It had been formed at least a century ago by a great accumulation of small and large fragments of sandstone, together with some sandy earth, rammed down level. It is probable that it was at first protected by being covered with turf. This terrace, judging from the number of castings on it, was rather unfavourable for the existence of worms, in comparison with the neighbouring fields and an upper terrace. It was indeed surprising that as many worms could live here as were seen; for on digging a hole in this terrace, the black vegetable mould together with the turf was only four inches in thickness, beneath which lay the level surface of light-coloured sandy soil, with many fragments of sandstone. Before any castings were collected all the previously existing ones were carefully removed. The last day's collection was on October 14th, 1871. The castings were then well dried before a fire; and they weighed exactly 3½ lbs. This would give for an acre of similar land 7·56 tons of dry earth annually ejected by worms.

The second square was marked on unenclosed common land, at a height of about 700 ft. above the sea, at some little distance from Leith Hill Tower. The surface was clothed with short, fine turf, and had never been disturbed by the hand of man. The spot selected appeared neither particularly favourable nor the reverse for worms; but I have often noticed that castings are especially abundant on common land, and this may, perhaps, be attributed to the poorness of the soil. The vegetable mould was here between three and four inches in thickness. As this spot was at some distance from the house where the lady lived, the castings were not collected at such short intervals of time as those on the terrace; consequently the loss of fine earth during rainy weather must have been greater in this than in the last case. The castings moreover were more sandy, and in collecting them during dry weather they sometimes crumbled into dust, and much was thus lost. Therefore it is certain that the worms brought up to the surface considerably more earth than that which was collected. The last collection was made on October 27th, 1871; i.e., 367 days after the square had been marked out and the surface cleared of all pre-existing castings. The collected castings, after being well dried, weighed 7·453 pounds; and this would give, for an acre of the same kind of land, 16·1 tons of annually ejected dry earth.

. . .

The thickness of the layer of mould, which castings ejected during a year would form if uniformly spread out.—As we know from the two last cases . . . the weight of the dried castings ejected by worms during a year on a square yard of surface, I wished to learn how thick a layer of ordinary mould this amount would form if spread uniformly over a square yard. The dry castings were therefore broken into small particles, and whilst being placed in a measure were well shaken and pressed down. Those collected on the Terrace amounted to 124·77 cubic inches; and this amount, if spread out over a square yard, would make a layer ·09612 inch in thickness. Those collected on the Common amounted to 197·56 cubic inches, and would make a similar layer ·1524 inch in thickness.

These thicknesses must, however, be corrected, for the triturated castings, after being well shaken down and pressed, did not make nearly so compact a mass as vegetable mould, though each separate particle was very compact. Yet mould is far from being compact, as is shown by the number of air-bubbles which rise up when the surface is flooded with water. It is moreover penetrated by many fine roots. To ascertain approximately by how much ordinary vegetable mould would be increased in bulk by being broken up into small particles and then dried, a thin oblong block of somewhat argillaceous mould (with the turf pared off) was measured before being broken up, was well dried and again measured. The drying caused it to shrink by ⅐ of its original bulk, judging from exterior measurements alone. It was then triturated and partly reduced to powder, in the same manner as the castings had been treated, and its bulk now exceeded (notwithstanding shrinkage from drying) by ¹⁄₁₆ that of the original block of damp mould. Therefore the above calculated thickness of the layer, formed by the castings from the Terrace, after being damped and spread over a square yard, would have to be reduced by ¹⁄₁₆; and this will reduce the layer to ·09 of an inch, so that a layer ·9 inch in thickness would be formed in the course of ten years. On the same principle the castings from the Common would make in the course of a single year a layer ·1429 inch, or in the course of 10 years 1·429 inch, in thickness. We may say in round numbers that the thickness in the former case would amount to nearly 1 inch, and in the second case to nearly 1½ inch in 10 years.

Burial of ancient buildings

Silchester, Hampshire. —The ruins of this small Roman town have been better preserved than any other remains of the kind in England. A broken wall, in most parts from 15 to 18 feet in height and about 1½ mile in compass, now surrounds a space of about 100 acres of cultivated land, on which a farm-house and a church stand. Formerly, when the weather was dry, the lines of the buried walls could be traced by the appearance of the crops; and recently very extensive excavations have been undertaken by the Duke of Wellington, under the superintendence of the late Rev. J. G. Joyce, by which means many large buildings have been discovered. Mr. Joyce made careful coloured sections, and measured the thickness of each bed of rubbish, whilst the excavations were in progress; and he has had the kindness to send me copies of several of them. When my sons Francis and Horace visited these ruins, he accompanied them, and added his notes to theirs.

Mr. Joyce estimates that the town was inhabited by the Romans for about three centuries; and no doubt much matter must have accumulated within the walls during this long period. It appears to have been destroyed by fire, and most of the stones used in the buildings have since been carried away. These circumstances are unfavourable for ascertaining the part which worms have played in the burial of the ruins; but as careful sections of the rubbish overlying an ancient town have seldom or never before been made in England, I will give copies of the most characteristic portions of some of those made by Mr. Joyce.[2] They are of too great length to be here introduced entire.

An east and west section, 30 ft. in length, was made across a room in the Basilica, now called the Hall of the Merchants [Fig. 10.2]. The hard concrete floor, still covered here and there with tesseræ, was found at 3 ft. beneath the surface of the field, which was here level. On the floor there were two large piles of charred wood, one alone of which is shown in the part of the section here given. This pile was covered by a thin white layer of decayed stucco or plaster, above which was a mass, presenting a singularly disturbed appearance, of broken tiles, mortar, rubbish and fine gravel, together 27 inches in thickness. Mr. Joyce believes that the gravel was used in making the mortar or concrete, which has since decayed, some of the lime probably having been dissolved. The disturbed state of the rubbish may have been due to its having been searched for building stones. This bed was capped by fine vegetable mould, 9 inches in thickness. From these facts we may conclude that the Hall was burnt down, and that much rubbish fell on the floor, through and from which the worms slowly brought up the mould, now forming the surface of the level field.

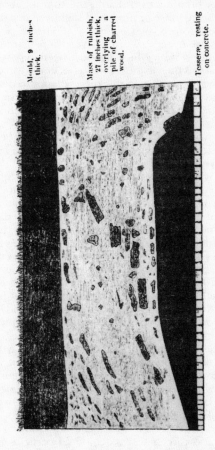

Mould, 9 inches thick.

Mass of rubbish, 27 inches thick, overlying a pile of charred wood.

Tesseræ, resting on concrete.

Figure 10.2 The basilica at Silchester
Section within a room in the Basilica at Silchester. Scale 1/18.

. . .

Turning now to the points which more immediately concern us. Worm-castings were observed on the floors of several of the rooms, in one of which the tesselation was unusually perfect. The tesseræ here consisted of little cubes of hard sandstone of about 1 inch, several of which were loose or projected slightly above the general level. One or occasionally two open worm-burrows were found beneath all the loose tesseræ. Worms have also penetrated the old walls of these ruins. A wall, which had just been exposed to view during the excavations then in progress, was examined; it was built of large flints, and was 18 inches in thickness. It appeared sound, but when the soil was removed from beneath, the mortar in the lower part was found to be so much decayed that the flints fell apart from their own weight. Here, in the middle of the wall, at a depth of 29 inches beneath the old floor and of 49½ inches beneath the surface of the field, a living worm was found, and the mortar was penetrated by several burrows.

A second wall was exposed to view for the first time, and an open burrow was seen on its broken summit. By separating the flints this burrow was traced far down in the interior of the wall, but as some of the flints cohered firmly, the whole mass was disturbed in pulling down the wall, and the burrow could not be traced to the bottom. The foundations of a third wall, which appeared quite sound, lay at a depth of 4 feet beneath one of the floors, and of course at a considerably greater depth beneath the level of the ground. A large flint was wrenched out of the wall at about a foot from the base, and this required much force, as the mortar was sound; but behind the flint in the middle of the wall, the mortar was friable, and here there were worm-burrows. Mr. Joyce and my sons were surprised at the blackness of the mortar in this and in several other cases, and at the presence of mould in the interior of the walls. Some may have been placed there by the old builders instead of mortar; but we should remember that worms line their burrows with black humus. Moreover open spaces would almost certainly have been occasionally left between the large irregular flints; and these spaces, we may feel sure, would be filled up by the worms with their castings, as soon as they were able to penetrate the wall. Rain-water, oozing down the burrows would also carry fine dark-coloured particles into every crevice. Mr. Joyce was at first very sceptical about the amount of work which I attributed to worms; but he ends his notes with reference to the last-mentioned wall by saying, "This case caused me more surprise and brought more conviction to me than any other. I should have said, and did say, that it was quite impossible such a wall could have been penetrated by earth-worms."

North.   Horizontal line.   South.

Figure 10.3 A subsided floor at Silchester
Section of the subsided floor of a room, paved with tesseræ, at Silchester.
Scale 1/40.

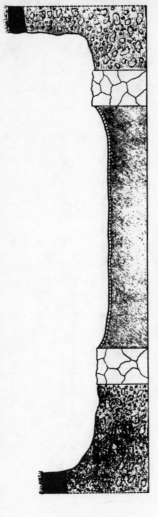

North.

Figure 10.4 A subsided floor at Silchester

A north and south section through the subsided floor of a corridor, paved with
tesserae. Outside the broken-down bounding walls, the excavated ground on each
side is shown for a short space. Nature of the ground beneath the tesserae
unknown. Scale 1/36.

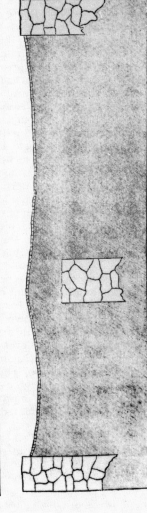

South.   Horizontal line.   North.

Figure 10.5   A subsided floor at Silchester.
Section through the subsided floor, paved with tesserae, and of the broken-down bounding walls of a room at Silchester, which had been formerly enlarged, with the foundations of the old wall left buried. Scale 1/40.

In almost all the rooms the pavement has sunk considerably, especially towards the middle; and this is shown in the three previous sections. The measurements were made by stretching a string tightly and horizontally over the floor. The section, [Fig. 10.3], was taken from north to south across a room, 18 feet 4 inches in length, with a nearly perfect pavement, next to the "Red Wooden Hut." In the northern half, the subsidence amounted to 5¾ inches beneath the level of the floor as it now stands close to the walls; and it was greater in the northern than in the southern half; but, according to Mr. Joyce, the entire pavement has obviously subsided. In several places, the tesseræ appeared as if drawn a little away from the walls; whilst in other places they were still in close contact with them.

In [Fig. 10.4], we see a section across the paved floor of the southern corridor or ambulatory of a quadrangle, in an excavation made near "The Spring." The floor is 7 feet 9 inches wide, and the broken-down walls now project only ¾ of an inch above its level. The field, which was in pasture, here sloped from north to south, at an angle of 3°40′. The nature of the ground on each side of the corridor is shown in the section. It consisted of earth full of stones and other débris, capped with dark vegetable mould which was thicker on the lower or southern than on the northern side. The pavement was nearly level along lines parallel to the side-walls, but had sunk in the middle as much as 7¾ inches.

A small room at no great distance from that represented in [Fig. 10.3], had been enlarged by the Roman occupier on the southern side, by an addition of 5 feet 4 inches in breadth. For this purpose the southern wall of the house had been pulled down, but the foundations of the old wall had been left buried at a little depth beneath the pavement of the enlarged room. Mr. Joyce believes that this buried wall must have been built before the reign of Claudius II., who died 270, A.D. We see in the accompanying section, [Fig. 10.5], that the tesselated pavement has subsided to a less degree over the buried wall than elsewhere; so that a slight convexity or protuberance here stretched in a straight line across the room. This led to a hole being dug, and the buried wall was thus discovered.

We see in these three sections, and in several others not given, that the old pavements have sunk or sagged considerably. Mr. Joyce formerly attributed this sinking solely to the slow settling of the ground. That there has been some settling is highly probable, and it may be seen in section 15 that the pavement for a width of 5 feet over the southern enlargement of the room, which must have been built on fresh ground, has sunk a little more than on the old northern side. But this sinking may possibly have had no connection with the enlargement of the room, for in [Fig. 10.3], one half of the pavement has subsided more than the other half without any assignable cause. In a bricked passage to Mr. Joyce's own house, laid down only about six years ago, the same kind of sinking has occurred as in the ancient buildings. Nevertheless it does not appear probable that the whole amount of sinking can be thus accounted for. The Roman builders excavated the ground to an unusual depth for the foundations of their walls, which were thick and solid; it is therefore hardly credible that they should have been careless about the solidity of the bed on which their tesselated and often ornamented pavements were laid. The sinking must, as it appears to me, be attributed in chief part to the pavement having been undermined by worms, which we know are still at work. Even Mr. Joyce at last admitted that this could not have failed to have produced a considerable effect. Thus also the large quantity of fine mould overlying the pavements can be accounted for, the presence of which would otherwise be inexplicable. My sons noticed that in one room in which the pavement had sagged very little, there was an unusually small amount of overlying mould.

As the foundations of the walls generally lie at a considerable depth, they will either have not subsided at all through the undermining action of worms, or they will have subsided much less than the floor. This latter result would follow from worms not often working deep down beneath the foundations; but more especially from the walls not yielding when penetrated by worms, whereas the successively formed burrows in a mass of earth, equal to one of the walls in depth and thickness, would have collapsed many times since the desertion of the ruins, and would consequently have shrunk or subsided. As the walls cannot have sunk much or at all, the immediately adjoining pavement from adhering to them will have been prevented from subsiding; and thus the present curvature of the pavement is intelligible.

The circumstance which has surprised me most with respect to Silchester is that during the many centuries which have elapsed since the old buildings were deserted, the vegetable mould has not accumulated over them to a greater thickness than that here observed. In most places it is only about 9 inches in thickness, but in some places 12 or even more inches. . . . The land enclosed within the old walls is described as sloping slightly to the south; but there are parts which, according to Mr. Joyce, are nearly level, and it appears that the mould is here generally thicker than elsewhere. The surface slopes in other parts from west to east, and Mr. Joyce describes one floor as covered at the western end by rubbish and mould to a thickness of 28½ inches, and at the eastern end by a thickness of only 11½ inches. A very slight slope suffices to cause recent castings to flow downwards during heavy rain, and thus much earth will ultimately reach the neighbouring rills and streams and be carried away. By this means, the absence of very thick beds of mould over these ancient ruins may, as I believe, be explained. Moreover most of the land here has long been ploughed, and this would greatly aid the washing away of the finer earth during rainy weather.

The nature of the beds immediately beneath the vegetable mould in some of the sections is rather perplexing. We see, for instance, in the section of an excavation in a grass meadow [Fig. 10.4], which sloped from north to south at an angle of 3°40', that the mould on the upper side is only six inches and on the lower side nine inches in thickness. But this mould lies on a mass (25½ inches in thickness on the upper side) "of dark brown mould," as described by Mr. Joyce, "thickly interspersed with small pebbles and bits of tiles, which present a corroded or worn appearance." The state of this dark-coloured earth is like that of a field which has long been ploughed, for the earth thus becomes intermingled with stones and fragments of all kinds which have been much exposed to the weather. If during the course of many centuries this grass meadow and the other now cultivated fields have been at times ploughed, and at other times left as pasture, the nature of the ground in the above section is rendered intelligible. For worms will continually have brought up fine earth from below, which will have been stirred up by the plough whenever the land was cultivated. But after a time a greater thickness of fine earth will thus have been accumulated than could be reached by the plough; and a bed like the 25½-inch mass, in [Fig. 10.4], will have been formed beneath the superficial mould, which latter will have been brought to the surface within more recent times, and have been well sifted by the worms.

. . .

Conclusion

Worms have played a more important part in the history of the world than most persons would at first suppose. In almost all humid countries they are extraordinarily numerous, and for their size possess great muscular power. In many parts of England a weight of more than ten tons (10,516 kilogrammes) of dry earth annually passes through their bodies and is brought to the surface on each acre of land; so that the whole superficial bed of vegetable mould passes through their bodies in the course of every few years. From the collapsing of the old burrows the mould is in constant though slow movement, and the particles composing it are thus rubbed together. By these means fresh surfaces are continually exposed to the action of the carbonic acid in the soil, and of the humus-acids which appear to be still more efficient in the decomposition of rocks. The generation of the humus-acids is probably hastened during the digestion of the many half-decayed leaves which worms consume. Thus the particles of earth, forming the superficial mould, are subjected to conditions eminently favourable for their decomposition and disintegration. Moreover, the particles of the softer rocks suffer some amount of mechanical trituration in the muscular gizzards of worms, in which small stones serve as mill-stones.

The finely levigated castings, when brought to the surface in a moist condition, flow during rainy weather down any moderate slope; and the smaller particles are washed far down even a gently inclined surface. Castings when dry often crumble into small pellets and these are apt to roll down any sloping surface. Where the land is quite level and is covered with herbage, and where the climate is humid so that much dust cannot be blown away, it appears at first sight impossible that there should be any appreciable amount of subaerial denudation; but worm-castings are blown, especially whilst moist and viscid, in one uniform direction by the prevalent winds which are accompanied by rain. By these several means the superficial mould is prevented from accumulating to a great thickness; and a thick bed of mould checks in many ways the disintegration of the underlying rocks and fragments of rock.

The removal of worm castings by the above means leads to results which are far from insignificant. It has been shown that a layer of earth, ·2 of an inch in thickness, is in many places annually brought to the surface per acre; and if a small part of this amount flows, or rolls, or is washed, even for a short distance down every inclined surface, or is repeatedly blown in one direction, a great effect will be produced in the course of ages. It was found by measurements and calculations that on a surface with a mean inclination of 9°26′, 2·4 cubic inches of earth which had been ejected by worms crossed, in the course of a year, a horizontal line one yard in length; so that 240 cubic inches would cross a line 100 yards in length. This latter amount in a damp state would weigh 11½ pounds. Thus a considerable weight of earth is continually moving down each side of every valley, and will in time reach its bed. Finally this earth will be transported by the streams flowing in the valleys into the ocean, the great receptacle for all matter denuded from the land. It is known from the amount of sediment annually delivered into the sea by the Mississippi, that its enormous drainage-area must on an average be lowered ·00263 of an inch each year; and this would suffice in four and half million years to lower the whole drainage-area to the level of the sea-shore. So that, if a small fraction of the layer of fine earth, ·2 of an inch in thickness, which is annually brought to the surface by worms, is carried away, a great result cannot fail to be produced within a period which no geologist considers extremely long.

Archæologists ought to be grateful to worms, as they protect and preserve for an indefinitely long period every object, not liable to decay, which is dropped on the surface of the land, by burying it beneath their castings. Thus, also, many elegant and curious tesselated pavements and other ancient remains have been preserved; though no doubt the worms have in these cases been largely aided by earth washed and blown from the adjoining land, especially when cultivated. The old tesselated pavements have, however, often suffered by having subsided unequally from being unequally undermined by the worms. Even old massive walls may be undermined and subside; and no building is in this respect safe, unless the foundations lie 6 or 7 feet beneath the surface, at a depth at which worms cannot work. It is probable that many monoliths and some old walls have fallen down from having been undermined by worms.

Worms prepare the ground in an excellent manner for the growth of fibrous-rooted plants and for seedlings of all kinds. They periodically expose the mould to the air, and sift it so that no stones larger than the particles which they can swallow are left in it. They mingle the whole intimately together, like a gardener who prepares fine soil for his choicest plants. In this state it is well fitted to retain moisture and to absorb all soluble substances, as well as for the process of nitrification. The bones of dead animals, the harder parts of insects, the shells of land-molluscs, leaves, twigs, &c., are before long all buried beneath the accumulated castings of worms, and are thus brought in a more or less decayed state within reach of the roots of plants. Worms likewise drag an infinite number of dead leaves and other parts of plants into their burrows, partly for the sake of plugging them up and partly as food.

  The leaves which are dragged into the burrows as food, after being torn into the finest shreds, partially digested, and saturated with the intestinal and urinary secretions, are commingled with much earth. This earth forms the dark coloured, rich humus which almost everywhere covers the surface of the land with a fairly well-defined layer or mantle. Von Hensen placed two worms in a vessel 18 inches in diameter, which was filled with sand, on which fallen leaves were strewed; and these were soon dragged into their burrows to a depth of 3 inches. After about 6 weeks an almost uniform layer of sand, a centimeter (·4 inch) in thickness, was converted into humus by having passed through the alimentary canals of these two worms. It is believed by some persons that worm-burrows, which often penetrate the ground almost perpendicularly to a depth of 5 or 6 feet, materially aid in its drainage; notwithstanding that the viscid castings piled over the mouths of the burrows prevent or check the rain-water directly entering them. They allow the air to penetrate deeply into the ground. They also greatly facilitate the downward passage of roots of moderate size; and these will be nourished by the humus with which the burrows are lined. Many seeds owe their germination to having been covered by castings; and others buried to a considerable depth beneath accumulated castings lie dormant, until at some future time they are accidentally uncovered and germinate.

Worms are poorly provided with sense-organs, for they cannot be said to see, although they can just distinguish between light and darkness; they are completely deaf, and have only a feeble power of smell; the sense of touch alone is well developed. They can therefore learn little about the outside world, and it is surprising that they should exhibit some skill in lining their burrows with their castings and with leaves, and in the case of some species in piling up their castings into tower-like constructions. But it is far more surprising that they should apparently exhibit some degree of intelligence instead of a mere blind instinctive impulse, in their manner of plugging up the mouths of their burrows. They act in nearly the same manner as would a man, who had to close a cylindrical tube with different kinds of leaves, petioles, triangles of paper, &c., for they commonly seize such objects by their pointed ends. But with thin objects a certain number are drawn in by their broader ends. They do not act in the same unvarying manner in all cases, as do most of the lower animals; for instance, they do not drag in leaves by their foot-stalks, unless the basal part of the blade is as narrow as the apex, or narrower than it.

When we behold a wide, turf-covered expanse, we should remember that its smoothness, on which so much of its beauty depends, is mainly due to all the inequalities having been slowly levelled by worms. It is a marvellous reflection that the whole of the superficial mould over any such expanse has passed, and will again pass, every few years through the bodies of worms. The plough is one of the most ancient and most valuable of man's inventions; but long before he existed the land was in fact regularly ploughed, and still continues to be thus ploughed by earth-worms. It may be doubted whether there are many other animals which have played so important a part in the history of the world, as have these lowly organised creatures. Some other animals, however, still more lowly organised, namely corals,

The action of worms

have done far more conspicuous work in having constructed innumerable reefs and islands in the great oceans; but these are almost confined to the tropical zones.

Notes

1 Leith Hill Place was the home of Josiah and Caroline Wedgwood, Darwin's cousin and sister.
2 I have extracted only the first of four.

List of Darwin's books

Journal of researches into the geology and natural history of the various countries visited by HMS Beagle, 1839. (Revised edn 1845: *A naturalist's voyage.*)

The zoology of the voyage of HMS Beagle. Edited and superintended by Charles Darwin. Pt I: *Fossil Mammalia*, Richard Owen, 1840. Pt II: *Mammalia*, George R. Waterhouse, 1839. Pt III; *Birds*, John Gould, 1841. Pt IV: *Fish*, Leonard Jenyns, 1842. Pt V: *Reptiles*, Thomas Bell, 1843.

The structure and distribution of coral reefs, 1842.

Geological observations on the volcanic islands, visited during the voyage of HMS Beagle, 1844.

Geological observations on South America, 1846.

A monograph of the fossil Lepadidae; or, pedunculated cirripedes of Great Britain, 1851.

A monograph of the sub-class Cirripedia, with figures of all the species. The Lepadidae: or, pedunculated cirripedes, 1851.

A monograph of the fossil Balanidae and Verrucidae of Great Britain, 1854.

A monograph of the sub-class Cirripedia, with figures of all the species. The Balanidae (or sessile cirripedes); the Verrucidae, &c, 1854.

On the origin of species by means of natural selection, or preservation of favoured races in the struggle for life, 1859.

On the various contrivances by which British and foreign orchids are fertilised by insects, and on the good effects of intercrossing, 1862.

The variation of animals and plants under domestication, 2 vols, 1868.

The descent of man, and selection in relation to sex, 2 vols, 1871.

The expression of the emotions in man and animals, 1872.

Insectivorous plants, 1875.

The movements and habits of climbing plants, 1875.

The effects of self and cross fertilization in the vegetable kingdom, 1876.

The different forms of flowers on plants of the same species, 1877.

The power of movement in plants, 1880.

The formation of vegetable mould, through the action of worms, 1881.

List of Darwin's books

Darwin's autobiography was first published in F. Darwin (ed.), *Life and letters of Charles Darwin*, 3 vols, 1887. It is now available separately, for instance in the edition of G. R. De Beer (Oxford University Press), and in the edition of his grand-daughter, Nora Barlow (W. W. Norton & Company, 1969).

The correspondence of Charles Darwin is being edited by F. Burkhardt, S. Smith, and others, and published by Cambridge University Press.

The Collected Papers of Charles Darwin, ed. P. H. Barrett, 2 vols, 1977. Chicago; University of Chicago Press.

Further reading

CHAPTER 1

Ghiselin, M. T. 1969. *The triumph of the Darwinian method.* Berkeley: University of California Press.

Kohn, D. 1980. Theories to work by: rejected theories, reproduction, and Darwin's path to natural selection. *Studies in History of Biology* 4, 67–170.

Ospovat, D. 1981. *The development of Darwin's theory.* Cambridge: Cambridge University Press.

CHAPTER 2

Yonge, C. M. 1958. Darwin and coral reefs. In *A century of Darwin*, S. A. Barnett (ed.), 245–66. London: Heinemann.

CHAPTER 3

Sulloway, F. J. 1984. Darwin and the Galapagos. *Biological Journal of the Linnean Society* 21, 29–59.

CHAPTER 4

Howard, J. 1982. *Darwin*. Oxford: Oxford University Press.

CHAPTER 8

Waterhouse, H. L. K. 1959. Cross- and self-fertilization in plants. In *Darwin's biological work*, P. R. Bell (ed.), 207–61. Cambridge: Cambridge University Press.

CHAPTER 9

Bell, P. R. 1959. The movement of plants in response to light. In *Darwin's biological work*, P. R. Bell (ed.), 1–49. Cambridge: Cambridge University Press.

Index